# Model-based Nonlinear Control of Aeroengines

Jiqiang Wang · Weicun Zhang · Zhongzhi Hu

# Model-based Nonlinear Control of Aeroengines

Springer

Jiqiang Wang
Jiangsu Province Key Laboratory
of Aerospace Power Systems
Nanjing University of Aeronautics
and Astronautics
Nanjing, Jiangsu, China

Prof. Weicun Zhang
School of Automation Science
and Electrical Engineering
University of Science and Technology
Beijing
Beijing, Beijing, China

Prof. Zhongzhi Hu
Institute for Aeroengine Research
Tsinghua University
Beijing, Beijing, China

ISBN 978-981-16-4452-8    ISBN 978-981-16-4453-5 (eBook)
https://doi.org/10.1007/978-981-16-4453-5



This Springer imprint is published by the registered company Springer Nature Singapore Pte Ltd.
The registered company address is: 152 Beach Road, #21-01/04 Gateway East, Singapore 189721, Singapore

# Preface

Model-based design has been one of the pivotal technologies to advanced control and health management of propulsion systems. It can fulfil advanced designs such as fault tolerant control, engine modes control, and direct thrust control. As a consequence, model-based design has become an important research area in the field of aero-engines due to its theoretical interests and engineering significance. Yet it is quickly realized that a wide range of concepts and methods exist for model-based design of aeroengines, e.g. safety assurance control, life extending control, reconfigurable control, fast response control, etc., to name just a few. And the list soon becomes too long to be exhausted. However, tracing the origins of model-based design of aero-engines leads to the "discovery" that all the approaches have the common feature of three ingredients: models, controls, and fault detection. Specifically, most of the advanced model-based designs require mathematical models with some mathematical optimizations for optimal controls; should fault tolerance function is expected, a mechanism for fault detection is necessitated.

This is a rough concept for model-based design of aeroengines. While acknowledging that model-based designs include models, controls, and fault detections, the design methodologies integrating models with controls are often abbreviated as model-based controls. Besides time delays and constraints handling, etc., one of the central issues in model-based controls is the tackling of nonlinearities. There are publications concerning with either nonlinear modelling or nonlinear controls, yet they are scattered throughout the literature. It is timely to provide a comprehensive summary on model-based nonlinear controls.

This monograph aims to develop systematic design methodologies to nonlinear control of aeroengines, covering a wide range from nonlinear real-time on-board modelling, nonlinear control designs to integrated modelling and control technology, etc. Indeed, aeroengine control has been challenging due to its both time-critical and safety-critical natures. For example, typical full authority digital engine controllers (FADECs) for implementing control algorithms are only 20 ms, *which frustrates many advanced optimization-based control design approaches*; meanwhile, many limits exist such as pressures and temperatures in compressors, turbines, and combustors; and this enforces stringent boundaries on control solutions, which can lead to catastrophic failures should these limits be violated. Consequently, conventional approaches to aeroengine controls are usually of trial-and-error nature, and this calls upon systematic and deep investigations into advanced control of aeroengine systems. This line of research thus focuses on:

(1) Modelling of aeroengine systems—both component-level and identification-based models have been extensively studied and compared;
(2) Advanced nonlinear control designs—set-point control, transient control, and limit-protection control approaches have all been investigated;

Consequently, a series of important results are obtained; and a systematic design methodology is developed which provides consistently enhanced performance over a large flight/operational envelope. The results are verified through extensive numerical simulations or hardware-in-the-loop (HIL) testing, and it is thus expected to provide useful guidance to practical engineering in aeroengine industry and research.

Finally, it is worth pointing out that *this is the first systematic attempt to conduct nonlinear control of aeroengines in China.* It is initiated by the authors with the project "Model-based Nonlinear Control of Gas Turbine Power Systems" (MBNC-GTPS). Over the past decade, the research project has been funded by quite a several funding bodies, and the first author would like to take this opportunity to express sincere thanks for the financial support by the Central Military Commission Foundation to Strengthen Program Technology Fund (No. 2019-JCJQ-JJ-347); Central Military Commission Special Fund for Defence Science, Technology and Innovation (No. 20-163-00-TS-009-096-01); State Foreign Affairs Bureau Fund for Introduction Plan of Foreign Experts (G20200010100); Aviation Science Fund of China-Xi'an 631 Research Institute (201919052001); Natural Science Foundation of Jiangsu Province Project (BK20140829); and Jiangsu Postdoctoral Science Foundation (1401017B). The second author would like to acknowledge the partial financial support from the National Natural Science Foundation of China (No. 61520106010, No. 61741302).

Nanjing, China — Jiqiang Wang
Beijing, China — Weicun Zhang
Beijing, China — Zhongzhi Hu

# Contents

# Symbols

| | |
|---|---|
| P | Pressure |
| $m_{in}$ | Inlet mass flow ($kgs^{-1}$) |
| $m_{out}$ | Outlet mass flow ($kgs^{-1}$) |
| $C_p$ | Specific heat at constant pressure |
| $C_a$ | Speed of sound |
| R | Gas constant |
| T | Temperature |
| Ma | Mach number |
| $P_s$ | Static pressure |
| H | Height (km) |
| □ | Bypass ratio |
| V | Volume |
| k | Flow coefficient |
| $\eta_{fr}$ | Fan conversion speed |
| $Q_i$ | Air flow/I section ($kgs^{-1}$) |
| $\delta$ | Dimensionless static pressure (=P/Pstd) |
| $\gamma$ | Ratio of specific heats |
| m | Total mass flow ($kgs^{-1}$) |
| $V_f$ | Fan velocity ($ms^{-1}$) |
| $A_n$ | Area of core nozzle ($m^2$) |
| $A_f$ | Area of fan nozzle ($m^2$) |
| $V_j$ | Jet velocity ($ms^{-1}$) |
| f | fuel to air ratio |
| $V_a$ | Air velocity ($ms^{-1}$) |
| $F_{bypass}$ | Bypass thrust (kN) |
| $F_{core}$ | Core thrust (kN) |
| $m_f$ | Mass fuel flow ($kgs^{-1}$) |
| F | Uninstalled thrust (kN) |
| sfc | Specific fuel consumption (mg/N-s) |
| $\pi$ | Pressure ratio |
| $N_l$ | Lower pressure rotor speed % |
| $N_h$ | High pressure rotor speed % |

| | |
|---|---|
| $\theta$ | Dimensionless temperature ratio (=T/Tstd) |
| $W$ | Workdone (J) |

# List of Figures

# List of Tables

# Chapter 1
# Introduction to Aeroengine Controls

Gas turbine engines and its related technologies represent one of the most efficient forms of propulsion and power generation. Their applications range from land-based power generation, ground-based vehicle propulsion, on-board power and propulsion sources for marine ships, to aircraft propulsion and auxiliary power systems. They represent arguably one of the most complex types of machines ever created by humans. A rough classification of gas turbine engines is shown below.

## 1.1 Introduction to Model-Based Designs

Referring to Fig. 1.1, jet propulsion engines are usually called aeroengines. And in this monograph, however, turbojet, turboshaft and turbofan engines will be particularly concerned. As airbreathing jet propulsion gas turbine engines, they are powering all of the helicopters and most of the long-range vehicles, including both commercial and military aircrafts. A typical structure of this type of engines is shown (Fig. 1.2).

While schematically simple, the engines are composed of over ten thousand components and parts. For such a complex system, it can be imagined that the control of gas turbine engines can be extremely challenging. The difficulties associated with control system design can be seen as follows:

(1) Design is challenging as engine performance varies among engines and deterioration over time;
(2) Control implementations need have flexibility of adaptation (particularly to in-field modifications);
(3) Control system must handle engine monitoring functions;
(4) Multiple tasks/requirements, such as:
    - power management;
    - start logic;
    - takeoff control (maximum thrust control);



J. Wang et al., *Model-based Nonlinear Control of Aeroengines*,
https://doi.org/10.1007/978-981-16-4453-5_1

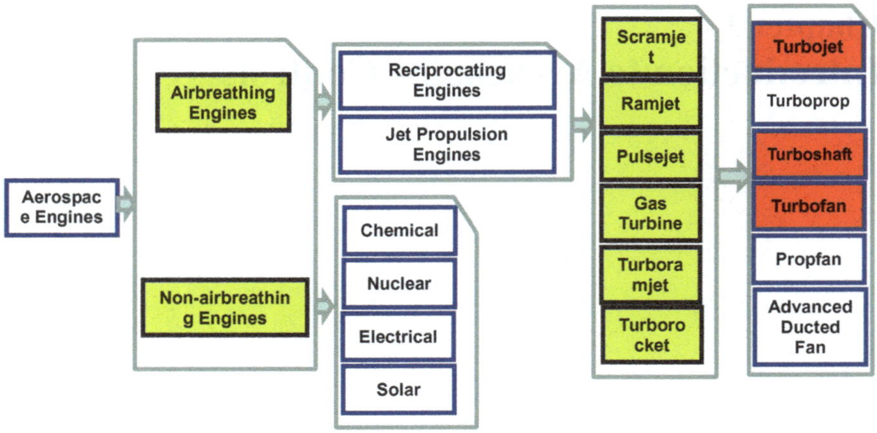


**Fig. 1.1** Classification of aerospace engines

**Fig. 1.2** Basic structure of turbofan engines

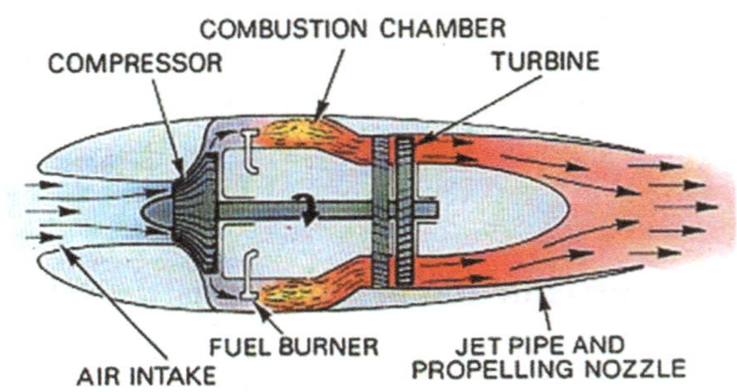


- climbing control (minimum time control);
- cruise control (minimum fuel flow control);
- acceleration and deceleration control;
- limit protection control;
- variable stator vane control;
- variable bleeding value control;
- turbine bleeding valve control;
- active clearance and other active controls;
- anti-stall and surge control;
- engine monitoring;
- other "intelligent" requirements etc.

Besides the performance requirements, there are:

(5) Structural and installation requirements: FADEC must restricted by size, weight etc.;

(6) Extreme safety and reliability: engine control system is both safety–critical and time-critical;

(7) Maintainability requirements;
(8) Airworthiness regulations, e.g. FAA, CAAC, ICAO all have regulations specified on design, reliability and $CO_2$ emission etc.;
(9) Other requirements, e.g. thrust vectoring, manoeuvrability, supersonic etc. To make things even "worse", all the requirements above should be satisfied/maintained under:
(10) Harsh working conditions—while flight control systems can sit in cockpit with "comfortable" condition, engine control system is usually installed on fan case, experiencing harsh environmental conditions, e.g. temperature 130 °C, vibration, electromagnetism, humidity, lightening, frogs and hails etc.

Therefore, designing such a safety–critical and time-critical system is indeed challenging, particularly under limited computational resource, e.g. usually 20 ms period for engine FADEC ECU. Henceforth design of propulsion control system is a systems engineering, requiring a thorough understanding to the thermal-aero-dynamics of engines, as well as the practical requirements from industries and regulation agencies. The design procedures must follow the procedures from model based designs (MBD): requirement analysis → preliminary design → detailed design → numerical simulation → dry test → wet test → rig test → flight test etc. In cases any procedure is not fulfilled, an iteration process is exercised and this is the so-called V-flow. A generic V-flow can be referred in the following Fig. 1.3.

Comparing with conventional design process, MBD suppresses many shortcomings such as slow prototyping, heavy workload, long development period, error-prone in practically every procedure of design process:

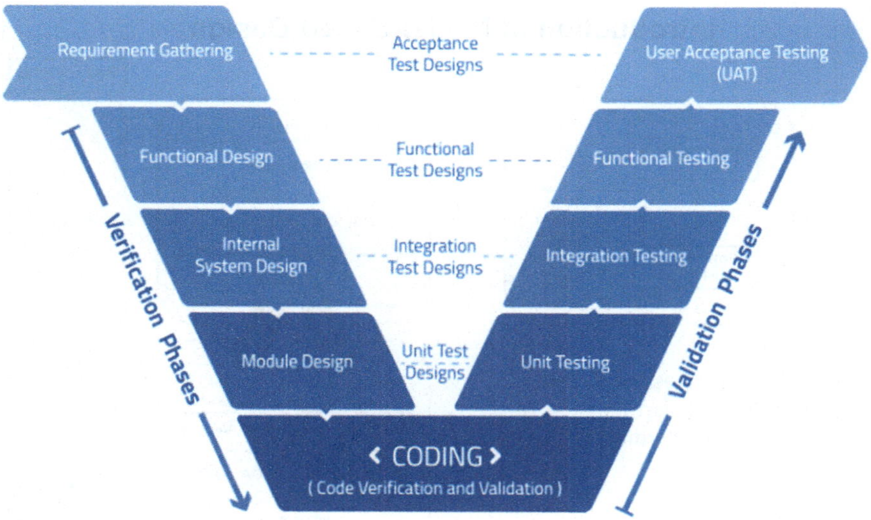


**Fig. 1.3** MBD covers the whole procedures in design

(1) For requirements analysis phase, conventional design utilizes hard-copied paper documentation, henceforth hard for fast iteration;
(2) For design phase, conventional design lacks of fast prototyping, resulting heavy investment of both capital and human resources;
(3) For implementation phase, conventional design still deploys in-site programming thus leading to long development cycle while producing errors;
(4) For validation phase, conventional design only "detects" errors during this phase, henceforth "forcing" a repetitive re-design of products.

Therefore MBD requires a design standard that covers the product development cycle. It can significantly accelerate the design period and move smoothly among research, design and implementation with less rework and less information loss (Fig. 1.4).

MBD has sparked many fundamental enabling technologies in propulsion system control (Fig. 1.5).

(1) **Fault tolerant control**: MBD can be utilized for fault detection, identification, isolation, diagnosis and even prognosis for sensors, actuators, components or even control systems, henceforth improving propulsion performance/reliability/safety/readiness rate, prolonging service life, reducing service and maintenance cost. Henceforth MBD is one of an enabling technology for engine health management (EHM) [1–4], which consists of functional components that provide advanced control and fault detection etc.
(2) **Performance seeking control**: for current operational condition, real time optimization technique can be used for optimizing a set of performance indices,

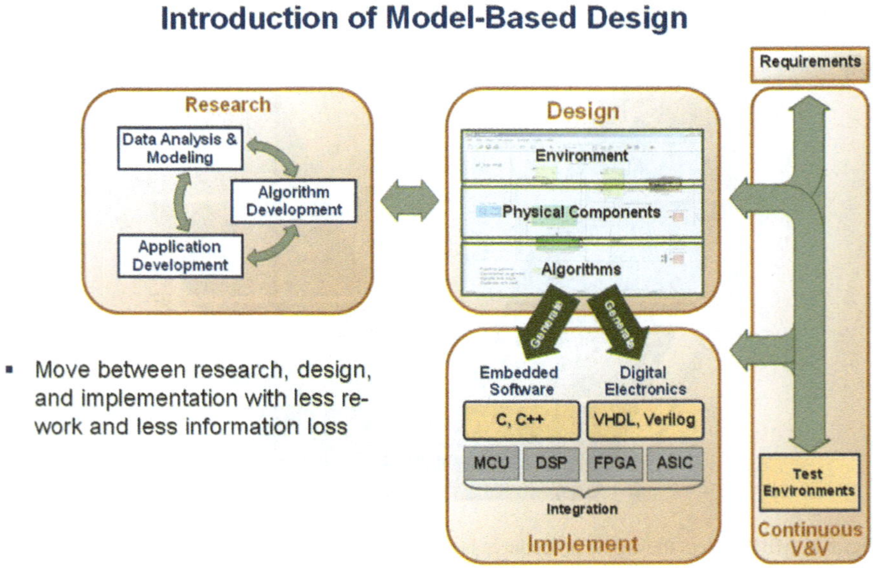


**Fig. 1.4** Introduction to MBD

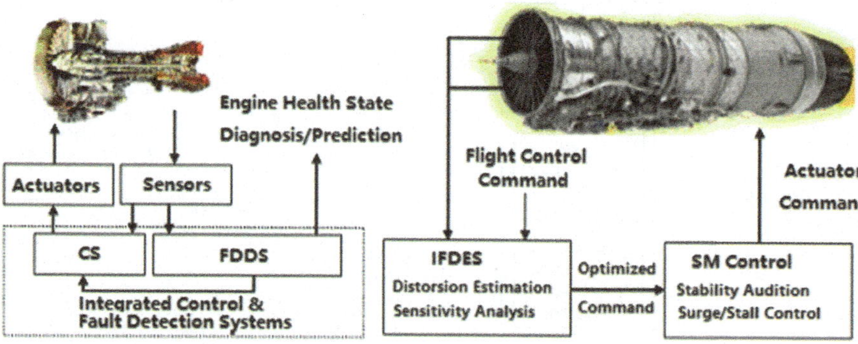


MBD for propulsion control & EHM High stability engine control

**Fig. 1.5** Application of model based design techniques in propulsion control

exploiting the potential of propulsion system, and thus leading to the assured flight operational quality. The optimization is exactly a model-based design process [5–8];

(3) **Direction propulsion control and high stability engine control**: real time models can be used for estimation of many variables that can be very difficult or prohibitively expensive to measure, e.g. turbine temperature, stall or surge margin, propulsion force, thus fulfilling the direct propulsion management [9, 10]; or they can be used to estimate the deterioration of propulsion performance and consequently recalibrate the regulation command, enabling high stability engine control [11–13];

(4) **Other advanced control**, e.g. life extending control [14, 15], active controls [16, 17] etc.

## 1.2 Introduction to Model-Based Controls

It has been illustrated that MBD takes care of the whole engine design cycle from a systems engineering perspective. Many enabling technologies such as those listed above are integrated in the process of preliminary design, detailed design and numerical simulation. Upon the detailed analysis and dissemination of system requirements, the structure and function of the propulsion system can be drafted. From a control system point of view, most of the model-based designs require mathematical models with some mathematical optimizations for optimal controls; should fault tolerance function is expected, a mechanism for fault detection is necessitated. This is a rough concept for model-based design of aeroengines. While acknowledging that model-based designs include models, controls, and fault detections, the design methodologies integrating models with controls are often abbreviated as model-based controls.

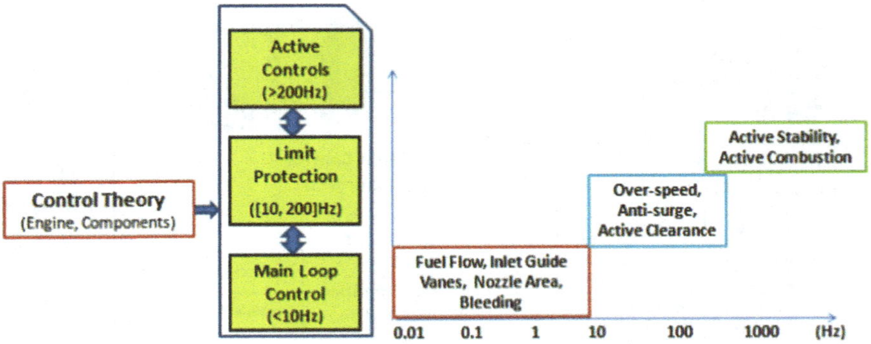


**Fig. 1.6** Engine control systems: frequency band of interest

Functionally, model-based controls, or the propulsion system in general, can be classified into three interacting control modes: primary control (fuel flow ratio control, guide vane control, nozzle area control, bleed valve control etc.), limit protection control (limit protection, anti-surge control, active clearance control), and active control (active stability control, active combustion control etc.). Fortunately, these three control modes have different frequency of interest and can thus be designed successively (refer to Fig. 1.6). However, primary control and limit protection control should work closely together for protecting the safe operation of the propulsion system.

As a consequence, they are combined together and are further divided functionally into three basic controls: set-point control, transient control and limit protection.

(1) Set-point control is to regulate the engine's performance near a desired operating condition, e.g. idle, cruise or takeoff;
(2) Transient control refers to the engine in transient operation when some or all of its performance variables are changing with time;
(3) During the transition, the engine should be protected from running beyond operating limits, e.g. physical limit of a shaft speed/turbine blades maximum temperature/maximum limit of burner or combustor pressure/surge or stall of compression system/other practically important parts.

Therefore, engine controls must provide the required control capability while respecting the corresponding limits. An example of the fuel control system operating envelope is given in Fig. 1.7. This functional design procedure usually work as follows: the flight envelope is divided into several regions and within each region, a linear model is obtained for the operational condition, e.g. small perturbation state space model [19, 20] or finite impulse response model [21]; then controllers are designed for each linear model using, e.g. PID control, LQR/LTR [22] or $H\infty$ optimal control [23–25]; finally the full flight envelope control is achieved through gain scheduling (usually scheduling variables are T1, P1, and the PI controller parameters are corrected based on the scheduling variables). This has been the standard

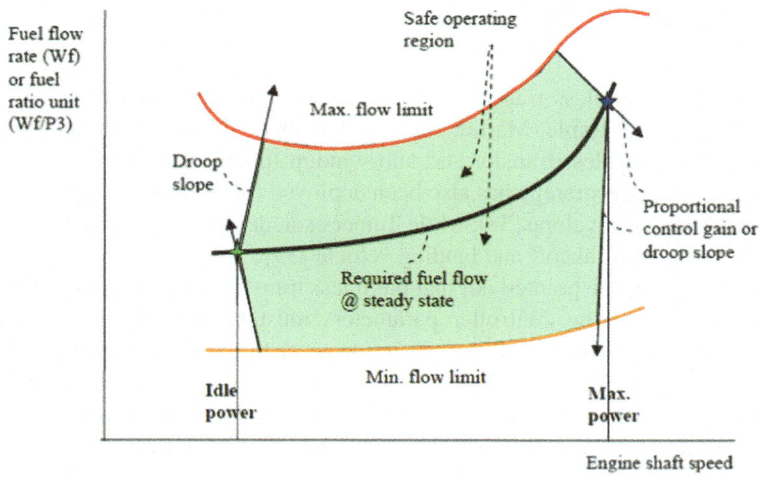


**Fig. 1.7** Engine fuel control system operating envelope (taken from [18])

practice in modern engine control designs and indeed, has resulted in fairly acceptable performance.

## 1.3 Model-Based Nonlinear Controls: State-of-the-Art

As an aeroengine is a complicated thermo-mechanical system, it can only be represented by a nonlinear process model. To capture the characteristics of the nonlinear aeroengine models, system identification approaches are usually adopted due to the real-time nature for control purposes. In this respect, neural networks [26, 27], generic algorithms [28], NARMAX [29], generalized describe function [30], Hammerstein-Wiener representation [31] methods are all utilized for nonlinear model identifications of aeroengines. Indeed, neural networks and system identification have been vast subjects and they open new avenues for aeroengine models [32, 33], and a systematic exploration is required that is covered in this book.

Nonlinear models of aeroengines surely necessitate the study of the nonlinear control techniques. While gain scheduling has been routinely utilized for nonlinear system control, however, it has also been suspected that gain scheduling requires switching among the controllers, and during switching, performance may seriously get deteriorated [34]. For this reason, although almost universally accepted, the designed controllers are in fact very conservative to provide "enough" stability margin. There has been necessitated for advanced controls to further reduce the conservatism and excavate the potential of the engine. Three measures are taken in the literature to handle this switching performance deterioration issue.

### *1.3.1 Bumpless Transfer*

The concept of bumpless transfer [35, 36] aims to prevent the detrimental effect of switching. For example, Martin, Bates and Wallace designed a gain scheduling controller with bumpless transfer and anti-windup functions for a turbofan engine model [37]; the same strategy has also been deployed by Frederick, Garg, and Adibhatla over the flight envelope [38], while Tumer et al. designed LQ bumpless transfer controller for a short takeoff and landing vehicle [39].

However, it must be pointed out that bumpless transfer design to gain-scheduling control implies that the controller parameters must be detuned for an already-completed-design. This is usually not desirable, since the parameters of the controller such as PID coefficients are obtained based on extensive testing and henceforth extremely expensive. It is thus not expected to be modified during the subsequent design procedures. This is the reason to introduce the concept of linear parametric varying in aeroengine designs.

### *1.3.2 Linear Parametric Varying*

Linear parameter varying control utilizes interpolation of linear model parameters for different operational conditions over the envelope, so that the performance also changes "contiguously" when switching among the different working conditions [40, 41]. Systematic design methods have been devised and applied using linear fractional transformations, see [42–45] and references therein. In the aaeroengine control industry, this linear parametric varying approach has been pursued extensively in SNECMA, and the interested readers can be referred to the series of articles [46–51]. The methodology has demonstrated to be of potential, however, the following challenges have also been identified: (1) the linear parameter varying model of aircraft engine is difficult to obtain; (2) control design is relatively difficult to perform due to the non-convexity of LMI; (3) the order of resulting controllers is relatively high. Therefore, linear parameter varying approach needs further investigation particularly for the benefit of on-board real-time models.

### *1.3.3 Full Flight Envelope Control*

While switching can cause performance deterioration, one natural idea is to avoid switching. This is the third research theme, which utilizes system identification or neural networks to obtain a feasible aeroengine model over the full flight envelope. Then well-established methods can be deployed, e.g. nonlinear predictive control [52] GE), multi-objective optimization and evolutionary algorithms [53, 54] (Rolls-Royce) and other intelligent control methods [55, 56].

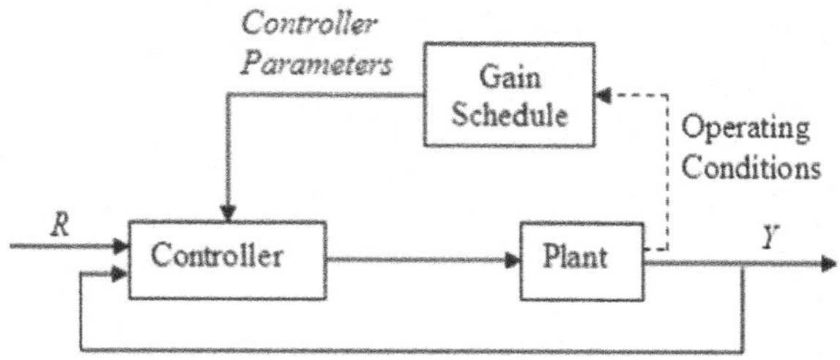


**Fig. 1.8** Gain scheduling adaptive control changes controller parameters in a pre-determined manner

## *1.3.4 Remark on Adaptive Control*

In fact, gain scheduling, linear parameter varying and model reference adaptive control techniques are all used to address plant variability, or change of aero-engine dynamics across the flight envelope. These control methods are fundamentally different from robust approaches in that controller parameters are not fixed, but undergo significant changes during system operation. Comparing with the $H_2/H_\infty$ control that is "static" and the parameters of the plant are considered to be fixed, henceforth the parameters of the controller to be designed. Both gain scheduling and model reference adaptive control are the two of the four most important forms of adaptive control. The other two main approaches to adaptive control are self-tuning regulators and adaptive dual control.

Gain scheduling is one of the simplest methods of adaptive controller design. Figure 1.8 shows the block diagram. Gain scheduling changes the controller parameters in a predefined manner to compensate the changes in the plant. Therefore, the controller can react quickly to changing conditions. However, it achieves this through an open-loop way: the scheduled controller changes parameters without monitoring closed-loop performance (this is indicated in Fig. 1.8 by dashed lines). As a consequence, gain scheduling can not be used if the plant dynamics or the disturbances are not known accurately. Nowadays gain scheduling has been an important ingredient in modern flight and engine control systems. In fact, the concepts of gain scheduling can be easily integrated with other control methods such as LQG and $H_\infty$ control: changing controller parameters according to a certain scheduling variable while designing controllers using LQG or $H_\infty$ techniques for a fixed scheduling value. Gain scheduling has become one of the most popular approaches to non-linear control design (see the survey article [40, 41]).

In model reference adaptive control (see Fig. 1.9), the output $Y$ of the plant is required to follow the model output $Y_M$ for a given reference $R$. That is, the desired performance is specified by a reference model. Hence the key problem with model reference adaptive control is to determine the adjustment mechanism so that a stable system, which brings the error to zero, is obtained [57]. There are many different methods for the determination of the adjustment mechanism, three of which are most popular: the gradient approach based on MIT rule; Lyapunov's second method and hyper stability design. In the gradient approach, the error $E = Y - Y_M$ and a cost function $J(\theta)$ is defined where $\theta$ is the parameter (or parameters) to be designed.

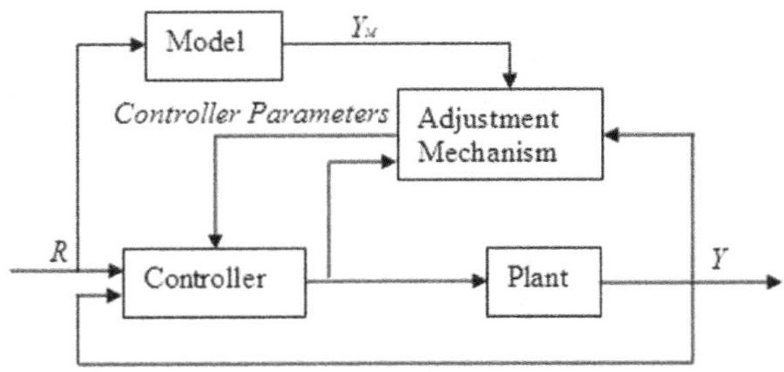


**Fig. 1.9** Block diagram of model reference adaptive control: the controller is such designed to make the performance $Y$ follow the desired model performance $Y_M$ for the same reference signal $R$

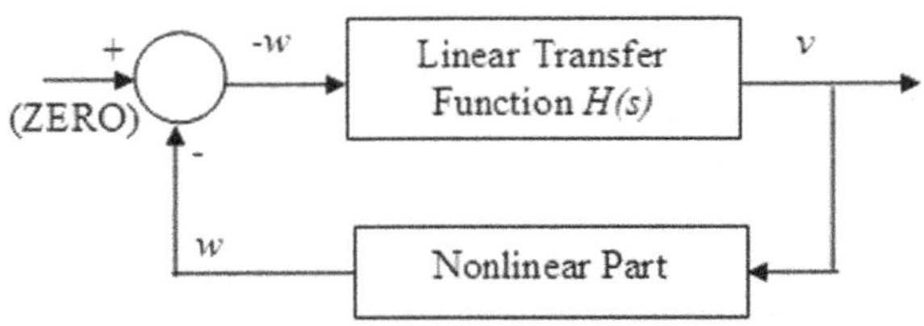


**Fig. 1.10** Hyper stability design is a direct method of stability analysis of nonlinear models, considered as linear and nonlinear parts [61]

Then the MIT rule states that the parameter (or parameters) is changed in the direction of the negative gradient of $J$ so that it is to be minimized:

$$\frac{d\theta}{dt} = -k\frac{\partial J}{\partial \theta} = -k\frac{\partial J}{\partial e}\frac{\partial e}{\partial \theta}(k > 0). \tag{1.1}$$

Therefore, if the parameter change is assumed to be slow then the item $\partial e/\partial \theta$ which is called sensitivity derivative, can be evaluated. Hence the controller adjustment mechanism can be determined by Eq. (1.1). Since the cost function can have many different forms (e.g. mean square error, absolute value error etc.) there are many alternatives to how the parameters are updated. The tuning of the adaptation gain $k$ is crucial in the gradient approach because the MIT rule does not guarantee convergence or stability. Lyapunov's second method is an alternative approach but with stability guarantee. In this approach, the adaptation mechanism for the controller is derived from the systems' Lyapunov functions. Therefore, the key problem is to define an appropriate Lyapunov function. However, intuition and experience have to be exercised to find a Lyapunov function, although this can be done easily for linear stable systems.[1] Many references on the control of dynamical systems can be made towards this purpose, e.g. [58–60]. Unlike the state-space view of the Lyapunov approach, hyper stability design employs an input–output view and decomposes the nonlinear model into a linear and a nonlinear part (Fig. 1.10).

Then the adjustment mechanism is to be derived by the utilization of Popov Integral Inequality (PII):

[1] For an asymptotically linear system $dx/dt = Ax$, it can be shown that for each symmetric $Q > 0$ there exists a unique $P > 0$ such that $A^T P + PA = -Q$. Then $V(x) = x^T Px$ is a Lyapunov function.

$$\int_{0}^{t_1} vw dt \geq -\lambda_0^2 \; \forall t_1 \geq 0. \tag{1.2}$$

Such designed controller will result in a hyper stable system if $H(s)$ is positive real, or asymptotic hyper stable if $H(s)$ is strictly positive real.

In all the main approaches to adaptive control (include self-tuning control), an implicit assumption is the certainty-equivalence (CE) principle: parameter estimation is separated from controller design. That is, the identified parameters are used in the controller as if they were the true values of the unknown plant. The uncertainty of the estimation is not taken into account. This may lead to poor control during tuning phase. Adaptive dual control (ADC) takes the task of treating the parameter uncertainties in a general optimization framework. However, ADC is too complicated for practical problems [57, 62]. Nevertheless, the development of ADC, particularly the bicriterial approach, raises the possibility of application. A reference can be made to the monograph [63] where a survey of dual control methods is presented. In particular the bicriterial synthesis method for dual controllers is delineated in great detail by the authors who are also the originators of this approach.

## 1.4 Contents and Contributions of the Monograph

From the discussions on the state-of-the-art of model-based nonlinear control and remarks on adaptive control in Sect. 1.3, it is readily identified that the current work has the following features:

(1) most of the model-based designs in aeroengine industry are based on linear models and henceforth linear control design methods such as $H_2/H_\infty$ optimization is utilized. This will involve with controller switching since a linear model is only valid over limited flight envelope, usually 3–5% of rotational speed. Thus, controller switching is inevitable for flight envelope control. Measures have to be taken to prevent performance deterioration during switching. However, as analysed in Sect. 1.3, both bumpless transfer and linear parametric varying approaches need to be further refined for applications, also see the discussions in the recent book [64].

(2) Among the model-based nonlinear control developments, most of the investigations utilize optimization-based methodology such as sequential quadratic programming, or multi-objective evolutionary algorithms. These will consume computational power, and indeed, the current engine electronic controller has very limited computational resource, with typical 20 ms sampling rate. This is far from running nonlinear optimization algorithms onboard. Thus, full-flight envelope control as reviewed in Sect. 1.3.3 would require new generation of engine control units to be proven.

(3) The issue of nonlinearity, or model-based nonlinear control has not been given enough considerations. Adaptive controls including the classical gain scheduling, and advanced ones as model reference nonlinear control and self-tuning control techniques aim to address the nonlinearity issue and full flight envelope control issue "simultaneously," yet this strategy has not been flight-tested and it is the authors' conviction that the objective is hard to achieve due to the current engine control requirements, e.g. harsh working conditions, limited computational powers, instability issues etc.

Thus, an aeroengine control engineer is confronted with the problem of frequent switching with model-based linear controls, on the one hand; on the other hand, model-based nonlinear control during full flight envelope is unachievable for on-board applications due to the limited computational resources. One way out is to take a compromise in the sense that by utilizing certain forms of nonlinear models, the number of switching can be reduced, while the corresponding nonlinear control designs do not rely heavily on optimization so that computationally sound.

To comprehend the situation, it is recalled that for aeroengine gain scheduling, the linear model is only valid during 3–5% of rotational speed, henceforth the corresponding controller needs to be switching accordingly [65–68]. Thus from idle state to takeoff power, there will be a dozen of switching, e.g. SPEY 202 engine requires 13 switchings even on ground, not to mention for flight envelope operations. On the other hand, if a component-level model feasible for full flight envelope is utilized, the corresponding controller needs to deploy complicated optimization algorithms that are not feasible for current electronics in engine control units. One interesting or even natural idea is to utilize a nonlinear model, and this model may be feasible during an extended flight envelope over linear counterpart, e.g. 10% or even 20% of rotational speed. This will substantially reduce the number of switchings; and at the same time, if the corresponding nonlinear control techniques are not computational extensive, such a compromise will be promising for current engineering practice in the industry and near future development.

Indeed, by employing nonlinearity, not only the model validity range is extended, but also the implementation complexity is reduced. These features are certainly welcomed by engine electronic control unit which has very limited computation power of only 20 ms sampling rate but needs ensure real-time control capability at the same time. Meanwhile, nonlinearity is an inherent property for aeroengines, and model-based nonlinear controls are thus necessitated from a physical perspective. It is thus expected that model-based nonlinear control with integrated nonlinear modelling and corresponding (real-time) nonlinear control techniques will open a new avenue for advanced control of aeroengines and engine health management in general. This is the strong motivation for proposing the developments in this monograph.

The monograph will initiate a systematic investigation upon model-based nonlinear control: aeroengine nonlinear onboard real-time modelling, model-based nonlinear design for high performance real-time control, and integrated nonlinear modelling and control over large flight envelope will be studied. These are fundamental issues in the field and important results are obtained. Specifically:

- ***For nonlinear onboard real-time modelling***, the following modelling techniques are proposed:

(1) Improved back-propagation neural network modelling;
(2) Hammerstein-Wiener representation modelling; and
(3) Nonlinear auto-regressive with exogenous input (NARX) modeling.

- ***For aeroengine set-point control***, the following nonlinear design methods are proposed:

(1) Generalized Gronwell-Bellman lemma approach for set-point control;
(2) Lyapunov-based set-point designs;
(3) Set-point PID control with Lyapunov functions;
(4) Joint set-point design with Gronwell-Bellman lemma and Lyapunov function;
(5) Nonlinear restricted set-point control, and
(6) Finite-time set-point control.

- ***For aeroengine transient control***, the following nonlinear design methods are introduced:

(1) Nonlinear generalized minimum variance transient control;
(2) Nonlinear predictive generalized minimum variance transient control;
(3) Nonlinear aeroengine transient control with online tuning; and
(4) Switching-based transient control.

- ***For aeroengine control integration***, the following optimization-based nonlinear design methods are introduced:

(1) Sequential quadratic optimization-based transient and limit protection control;
(2) Active set method for transient and limit protection control;
(3) Comparative study for optimization-based designs.

This finishes the introduction to the topics covered in this monograph, and focus will be on the fundamentals of the three basic functions and their designs will be delineated but with particular emphasis on the utilization of nonlinear design techniques. It is not intended to be compressive but to disseminate the state-of-art in the field of advanced control of aircraft engines. Therefore, the book will be based on the work that the authors have been undertaken during the past decades. Towards this purpose, the book is structured as follows:

Chapter 2 introduces the component-level modelling and system identification modelling as a self-contained introduction to aeroengine models. With this background, nonlinear advanced modelling techniques such as improved back-propagation neural network modelling, Hammerstein-Wiener representation modeling; and nonlinear auto-regressive with exogenous input (NARX) modeling approaches will be introduced subsequently.

Chapter 3 introduces the concept of set point control, and present a series of nonlinear control design methods such as generalized Gronwall-Bellman Lemma based control, Lyapunov function based methodologies, and finite time control.

Chapter 4 presents the acceleration and deceleration control design methods. Schedule-based transient control and acceleration-based control design methods are introduced and both are conventionally utilized in the current engine control systems. Then a series of newly introduced acceleration and deceleration methods such as nonlinear generalized minimum variance-based control, nonlinear predictive generalized minimum variance-based control, Hammerstein-Wiener model-based control, and nonlinear predictive generalized minimum variable control with a state-dependent representation will be introduced.

Chapter 5 takes a collection of the results in previous chapters and assembles them into an integral control system capable of controlling the engines over the flight envelope.

Finally, conclusions and discussions for further developments are given in Chapter 6.

## References

1. Link, C. 2005. Recent advancements in aircraft engine health management (EHM) technologies and recommendations for the next step, GT-2005–68625.
2. Curtis, A., and S. Positron. 2007. Life remaining prognostics for airframe structural components. *IEEE-4244-0525-1245.*
3. Guidelines for integration of engine monitoring functions with on-board aircraft systems. 1996. *SAE-AIR4061.*
4. Donald, L., G. Sanjay, and M. Venti. 2003. Propulsion control and health management (PCHM) technology for flight test on the C-17 T-1 aircraft. *NASA/TM-2004-213303.*
5. Nobbs, S.G., S.W. Jacobs, and D.J. Donahue. 1992. Development of the full-envelop performance seeking control algorithm. *AIAA-92–3745.*
6. Gilyard, G., and J. Orme. 1993. Performance-seeking control: program overview and future directions. *AIAA-93–3765-CP.*
7. Gilyard, G., and J. Orme. 1992. Subsonic flight test evaluation of a performance seeking control algorithm on an F-15 airplane. *AIAA-92-3743.*
8. Orme, J., and T. Conners. 1994. Supersonic flight test results of a performance seeking control algorithm on a NASA F-15 aircraft. *AIAA-94-3210.*
9. Monaco, J., and D. Ward, and A. Bateman. 2004. A retrofit architecture for model-based adaptive flight control. In *AIAA 1st Intelligent Systems Technical Conference* 20–22 Septe 2004, Chicago, Illinois. AIAA 2004-6281.
10. Litt, J., N. Shah, and T. Sowers. 2005. A demonstration of a retrofit architecture for intelligent control and diagnostics of a turbofan engine. *NASA/TM-2005-214019.*
11. Freeman, C., A.G. Wilson, I.J. Day, and M.A. Swinbanks. 1998. Experiments in active control of stall on an aeroengine gas turbine. *ASME Journal of Turbomachinery* 120: 637–647.
12. Orme, J., J. DeLaat, R. Southwick, G. Gallops, and P. Doane. 1998. Development and testing of a high stability engine control (HISTEC) system. *NASA/TM 1998-206562.*
13. Southwick, R., G. Gallops, L. Kerr, R. Kielb, M. Welsh, J. DeLaat, and J. Orme. High stability engine control (HISTEC) flight test results. *NASA/TM 1998-208655.*
14. Ray, A., and J. Caplin. 2001. Life extending control of aircraft: Trade-off between flight performance and structural durability. *The Aeronautical Journal* 104: 397–408.
15. Jaw, L., and J. Mattingly. 2009. *Aircraft Engine Controls: Design, System Analysis, and Health Monitoring.* Virginia: AIAA Inc.

16. Mattern, D., and L. Law. 1997. Experimental results of an active tip clearance control system for a centrifugal compressor. In *Proceedings of International Gas Turbine Congress and Exposition.*
17. Johnson, C., Y. Neumeier, D. Darling, S. Sattinger, M. Neumaier, and B. Zinn. 2001. Demonstration of active control of combustion instabilities on a full-scale gas turbine combustor. *ASME*. paper 2001-GT-0519.
18. Jaw, L.C., and S. Garg. Propulsion control technology development in the United States, NASA/TM-2005–213978.
19. Zhou, W., X. Shan, Z. Geng, and J. Huang. 2008. Establishment of state space model of turboshaft engine with self-optimized method. *Journal of Aerospace Power* 23 (12): 2314–2320.
20. Zheng, T., X. Wang, X. Luo, and Q. Li. 2005. Modified method of establishing the state space model of aeroengine. *Journal of Propulsion Technology* 26 (1): 46–49.
21. Zhao, L., D. Fan, and W. Shan. 2008. Single-spool turbofan engine model identification. *Journal of Propulsion Technology* 29 (6): 733–736.
22. Yang, G., J. Sun, H. Yao, J. Zang, A. Liu, X. Yi, and L. Zhang. 2006. Experimental verification of H∞/LTR method for aeroengine control systems. *Acta Aeronautica et Astronautica Sinica* 27 (5): 773–777.
23. Härefors, M. 1997. Application of H∞ robust control to the RM12 jet engine. *Control Engineering Practice* 5 (9): 1189–1201.
24. Frederick, D.K., S. Garg, and S. Adibhatla. 2000. Turbofan engine control design using robust multivariable control technologies. *IEEE Transactions on Control Systems Technology* 8 (6): 961–970.
25. Wang, H., Y. Guo, and G. Li. 2008. Aero-engine control using two-degrees-of-freedom H∞ approach. In *Proceedings of ISSCAA*, pp. 1–5.
26. Embrechts, M.J., A.L. Schweizerhof, M. Bushman, et al. 2000. Neural network modeling of turbofan parameters. *ASME paper 2000-GT-0036.*
27. Zhuo, G., J. Sun, and G. Yang. 2004. Aeroengine modeling based on wavelet neural network. *Journal Nanjing University of Aeronautics and Astronautics* 36: 728–731.
28. Liu, Y., H. Xing, and S. Huang. 2012. Adaptive simulation of gas turbine performance using improved genetic algorithm. *J Aerosp Power* 27: 695–700.
29. Mu, J., D. Rees, and G. Liu. 2005. Advanced controller design for aircraft gas turbine engines. *Control Engineering Practice* 13 (8): 1001–1015.
30. Lichtsinder, M., and Y. Levy. 2006. Jet engine model for control and real-time simulations. *ASME Journal of Engineering for Gas Turbines and Power* 128: 745–753.
31. Wang, J., Z. Ye, and Z. Hu. 2014. Onboard real time modeling of aircraft engines with a Hammerstein-Wiener representation. *Journal Aerosp Power* 29: 2499–2506.
32. Irwin, G.W., K. Warwich, and K.J. Hunt. Neural network applications in control. IET Digital Library.
33. Ljung, L. 1999. *System Identification: Theory for the User*. 2nd ed. Upper Saddle River, NJ: PTR Prentice Hall.
34. Wang, J., Z. Ye, and Z. Hu. 2012. Nonlinear control of aircraft engines using a generalized Gronwall-Bellman lemma approach. *ASME Journal of Engineering for Gas Turbines and Power* 134 (9): 094502.
35. Zheng, K., T. Basar, and J. Bentsman. 2009. H∞ bumpless transfer under controller uncertainty. *IEEE Transactions on Automatic Control* 54 (7): 1718–1723.
36. Bendtsen, J.D., J. Stoustrup, and K. Trangbak. 2003. Bumpless transfer between advanced controllers with application to power plant control. In *Proceedings of 42nd IEEE Conference on Decision and Control,* pp. 2059–2064.
37. Martin, S., I. Wallace, and D.G. Bates. 2008. Development and validation of a civil aircraft engine simulation model for advanced controller design. *ASME Journal of Engineering for Gas Turbines and Power,* 130 (5): 051601:1–15.
38. Frederick, D.K., S. Garg, and S. Adibhatla. 2000. Turbofan engine control design using robust multivariable control technologies, *IEEE Transactions on Control Systems Technology* 8 (6): 961–970.

39. Tumer, M., N. Aouf, D.G. Bates, I. Postlethwaite, and B. Boulet. 2002. A switching scheme for full-envelope control of a V/STOL aircraft using LQ bumpless transfer. In *Proceedings of 2002 IEEE International Conference on Control Applications*, pp. 120–125.
40. Rugh, W., and J. Shamma. 2000. Research on gain scheduling. *Automatica* 36: 1401–1425.
41. Leith, D., and W. Leithead. 2000. Survey of gain-scheduling analysis and design. *International Journal of Control* 73 (11): 1001–1025.
42. Apkarian, P., P. Gahinet, and G. Becker. 1995. Self-scheduled H∞ control of linear parameter-varying systems: A design example. *Automatica* 31: 1251–1261.
43. Wu, F., X. Yang, A. Packard, and G. Becker. 1996. Induced L2-norm control for LPV systems with bounded parameter variation rates. *International Journal of Robust and Nonlinear Control* 6: 983–998.
44. Stilwell, D., and W. Rugh. 2000. Stability preserving interpolation methods for the synthesis of gain scheduled controllers. *Automatica* 36: 665–671.
45. Chang, Y., and B. Rasmussen. 2008. Stable controller interpolation for LPV systems. *Proceedings of ACC* 2008: 3082–3087.
46. Balas, G. 2002. Linear parameter-varying control and its application to a turbofan engine. *International Journal of Robust and Nonlinear Control* 12 (9): 763–793.
47. Bruzelius, F. 2004. *Linear Parameter-varying Systems-an Approach to Gain Scheduling*. PhD Thesis, Chalmers University, Göteborg, Sweden.
48. Henrion, D., L. Reberga, J. Bernussou, and F. Vary. 2004. Linearization and identification of aircraft turbofan engine models. In *Proceedings of IFAC Symposium on Automatic Control in Aerospace*, St. Petersburg.
49. Reberga, L., D. Henrion, J. Bernussou, and F. Vary. 2005. LPV modeling of a turbofan engine. In *Proceedings of IFAC World Congress on Automatic Control in Aerospace*, Prague, Czech Republic.
50. Vary, F., and L. Reberga. 2005. Programming and computing tools for jet engine control design. In *Proceedings of IFAC World Congress on Automatic Control*, Prague, Czech Republic.
51. Gilbert, W., D. Henrion, J. Bernussou, and D. Boyer. 2010. Polynomial LPV synthesis applied to turbofan engines. *Control Engineering Practice* 18: 1077–1083.
52. Brunell, B.J., R.R. Bitmead, and A.J. Connolly. 2002. Nonlinear model predictive control of an aircrift gas turbine engine. In *Proceedings of 41st IEEE Conference on Decision and Control*, Las Vegas, USA.
53. Fleming, P.J., and R.C. Purshouse. 2002. Evolutionary algorithms in control systems engineering: A survey. *Control Engineering Practice* 10 (11): 1223–1241.
54. Lyantsev, O.D., T.V. Breikin, G.G. Kulikov, and V.Y. Arkov. 2003. On-line performance optimization of aero engine control system. *Automatica* 39: 2115–2121.
55. Yao, Y., and J. Sun. 2008. 2008, Aeroengine direct thrust control based on neural network inverse control. *Journal of Propulsion Technology* 29 (2): 249–252.
56. Qi, X., and D. Fan. 2005. Application of improved FSQP algorithm to turbofan engine nonlinear multivariable control. *Journal of Propulsion Technology* 26 (1): 58–61.
57. Åström, K.J., and B. Wittenmark. 1994. *Adaptive Control*. 2nd ed. Addison-Wesley Longman Publishing Co., Inc.,.
58. Shankar, S. 1999. *Nonlinear Systems: Analysis*. Stability and Control: Springer Verlag, New York.
59. Vidyasagar, M. 1993. *Nonlinear Systems Analysis*, 2nd ed. Englewood Cliffs, New Jersey: Prentice Hall.
60. Slotine, J.J.E., and W. Li. 1991. *Applied Nonlinear Control*. Englewood Cliffs, New Jersey: Prentice Hall.
61. Tokhi, M.O. 2004. *Adaptive*. Self-Tuning Control: Lecture Notes, University of Sheffield.
62. Isermann, R., K.H. Lachmann, and D. Matko 1992. *Adaptive Control Systems*, Prentice Hall.
63. Filatov, N.M., and H. Unbehauen. 2004. *Adaptive Dual Control: Theory and Applications*. Berlin Heidelberg: Springer.
64. Richter, H. 2012. Advanced control of turbofan engines, Springer Science-Business Media, Chapter 5 on Gain Scheduling and Adaptation.

65. Samar, R. 1995. *Robust Multi-Mode Control of High Performance Aeroengines*. PhD Thesis, the University of Leicester.
66. Sutton, A.E. 1992. The application of multivariable control to a turbofan engine, *Technical Report TMP1220*, Defence Research Agency, Aerospace Division, Farnborough, Hampshire, UK.
67. Greig, A.W.M. Multivariable powerplant control-proof of concept, *Technical Report DRA/AS/PTD/TR/94067/1*, Defence Research Agency, Farnborough, Hampshire, UK, 1994.
68. Eisa, S.A., and H.P. Tyler. 1986. Closed loop control of an afterburning F100 gas turbine engine. In *Proceedings of the American Control Conference*, Seattle, WA, USA, pp. 266–272.

# Chapter 2
# Aeroengine Nonlinear Modeling

Advanced control often assumes implicitly that an aeroengine model is available. Modeling for both engine and control systems plays an important role in engine research and development: accelerate schedules while reducing cost and risks. For example, on control system design phase, models can be used for validating the corresponding control functions; on controller software and hardware design phase, models are used in fast prototyping, hardware-in-loop simulation, and semi-physical testing; on control system implementation phase, on-board models can be used for advanced control and fault detection, e.g. sensor analytical redundancy, flight/propulsion integration, direct propulsion control etc.

Multiple types of models exist, thus there are different classification schemes. In terms of model types, it can be classified component level models and identification models, where the former is also called first principle models while the latter identification-based models; in terms of real time performance, there are real time models and non-real time models; in terms of model characteristics: linear models and nonlinear models; in terms of model property: dynamic models and steady state models etc. With research and development, new model types and methods will appear. These are summarized in Table 2.1.

## 2.1 Aeroengine Component-Level Models

Most of the engine models are developed from the fundamental physics of turbomachinery—component-level models. Indeed, all the model types are derived from component-level models and they are often regarded as a "digital twin" to a real engine for model-based designs. Structurally an engine mainly consists of inlet, fan, compressors, high-pressure turbines, low-pressure turbines, mixer, and nozzle (Fig. 2.1), while functionally the propulsion process works as follow: air flow enters inlet experiencing initial pressure increase. It passes through fan blades with further pressure increase before flow separation: one portion goes to bypass while the

 
J. Wang et al., *Model-based Nonlinear Control of Aeroengines*,
https://doi.org/10.1007/978-981-16-4453-5_2

**Table 2.1** Model types and classifications

| Model classification schemes | |
|---|---|
| Model types | Component-level models versus identification models |
| Real-time performance | Real-time models versus non-real-time models |
| Model characteristics | Linear models versus nonlinear models |
| Model properties | Dynamic models versus steady-state models |
| Use of purpose | Control-oriented models versus fault detection-oriented models |
| Use of field | Off-line models versus on-board models |
| Others | New model types and methods can appear with research and development |

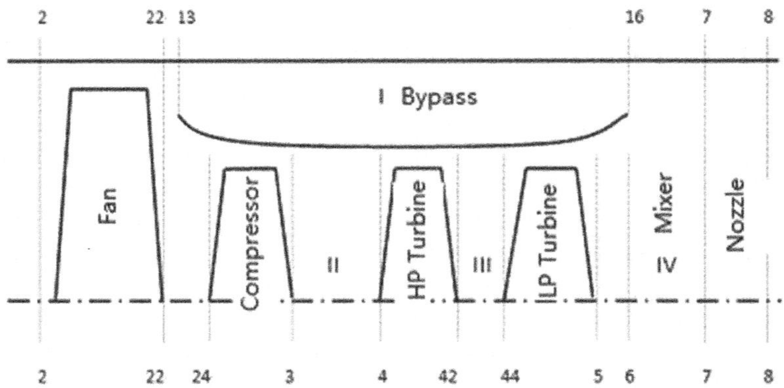


**Fig. 2.1** Typical engine components besides control systems and fuel delivery system

remaining enters into compressors for further mechanical supercharge. The supercharged air flow is propelled into burner, where it is mixed with fuel flow and experiences fuel injection combustion. The combustion-gas stream has a very high pressure and temperature (over 2000 K). This high temperature and high-pressure gas flows through the turbine expansion work, part of which transforms into mechanical work, providing driving power for fan blades, compressors, and other rotating accessories. As the expansion work produced by turbines exceeds that consumed by fan, compressors, and accessories, the gas flow at the exit of low-pressure turbines still possesses enormous heat and pressure energy. This part of energy will then be transformed into kinetic energy over the expansion process in the nozzle. Meanwhile the air flow passing through bypass duct is mixing with the gas flow, reducing the exit flow speed, increasing propulsion efficiency.

### 2.1.1 Aeroengine Components Dynamics

To design a control system that can recover engine performance by changing components' dynamics, the equations describing the properties of each component should be given. These are listed below [1, 2]:

***Inlet*** The engine intakes air flow from inlet, thus the inlet should provide appropriate mass flow to match the downstream components, particularly to fan blades. The inlet as an agent has its own regulation properties. For given Mach number $Ma$ and height $H$, the inputs and outputs of inlet can be calculated as below:

For environmental temperature $T_{s0}$:

$$T_{s0} = \begin{cases} 288.15 - 0.0065H, & H \leq 11000 \\ 216.5, & H > 11000 \end{cases} \tag{2.1}$$

For environmental pressure $P_{s0}$:

$$P_{s0} = \begin{cases} 101{,}325(1 - 0.225577 \times 10^{-4}H)^{5.25588}, & H \leq 11000 \\ 22{,}632\, e^{\frac{11{,}000-H}{6328}}, & H > 11{,}000 \end{cases} \tag{2.2}$$

For Inlet total temperature $T_{t1}$:

$$T_{t1} = T_{s0}\left(1 + \frac{\kappa - 1}{2}Ma^2\right) \tag{2.3}$$

where: $\kappa$ is the heat capacity ratio.

For Inlet total pressure $P_{t1}$:

$$P_{t1} = P_{s0}\left(1 + \frac{\kappa - 1}{2}Ma^2\right)^{\frac{\kappa}{\kappa-1}} \tag{2.4}$$

For Inlet inlet flow speed:

$$v_0 = Ma\sqrt{\kappa R T_{S0}} \tag{2.5}$$

Now assume the Inlet pressure recovery coefficient:

$$\begin{cases} \sigma_{in} = 1, & Ma \leq 1.0 \\ \sigma_{in} = 1 - 0.075(Ma - 1)^{1.35}, & Ma > 1.0 \end{cases} \tag{2.6}$$

Then the fan inlet total pressure $P_{t2}$:

$$P_{t2} = P_{t1}\sigma_{in} \tag{2.7}$$

Further assume fan inlet total temperature $T_{t2}$ equals to Inlet total temperature:

$$T_{t2} = T_{t1} \tag{2.8}$$

***Fan*** Air flow passes through fan blades and its pressure is increased with the work exerted by fan blades. Usually, the properties of fan are interpolated through fan characteristic table, and the corresponding operation is dealt with through fan corrected speed defined by:

$$n_{L,cor} = (n_L/\sqrt{T_{t2}})/(n_{Ld}/\sqrt{T_{t2d}}) \tag{2.9}$$

where: $T_{t2}$ is the fan inlet total temperature, $n_{Ld}$ is the low-pressure shaft speed at the design point, $T_{t2d}$ is the fan inlet total temperature at the design point. Through the fan characteristic table or plot, fan corrected flow $W_{af,cor}$ and efficiency $\eta_f$ can be "read out" as a lookup table in terms of fan corrected speed $n_{L,cor}$ and fan pressure ratio $\pi_f$.

Then fan inlet flow $W_{a2}$ is obtained as:

$$W_{a2} = W_{af,cor}\frac{P_2}{101{,}325}\sqrt{\frac{288.15}{T_{t2}}} \tag{2.10}$$

Fan outlet total temperature $T_{t22}$ is:

$$T_{t22} = T_{t2}\left[\left(\pi_f^{\frac{\kappa-1}{\kappa}} - 1\right)/\eta_f + 1\right] \tag{2.11}$$

Thus, fan outlet flow $W_{a22}$ is:

$$W_{a22}=W_{a2} \tag{2.12}$$

The power of the fan blades is also obtained:

$$N_f = W_{a2} \times c_p(T_{t22} - T_{t2}) \tag{2.13}$$

***Compressors*** The compressors work on the air flow coming through fan blades, increasing further air pressure, in order to facilitate the combustion process in the burner. Their physics are similar with the fan as follows:

The compressor corrected speed $n_{H,cor}$ is defined as:

$$n_{H,cor} = (n_H/\sqrt{T_{t24}})/(n_{Hd}/\sqrt{T_{t24d}}) \tag{2.14}$$

where: $T_{t24}$ is compressor inlet total temperature, $n_{Hd}$ is the high-pressure shaft speed at the design point, $T_{t24d}$ is the compressor inlet total temperature at design point. Similarly, the compressor corrected flow $W_{ac,cor}$ and compressor efficiency $\eta_c$ is obtained through the compressor characteristic table in terms of compressor corrected speed $n_{H,cor}$ and compressor pressure ratio $\pi_c$.

The compressor inlet flow $W_{a24}$ is:

$$W_{a24} = W_{ac,cor} \frac{P_{t24}}{101{,}325} \sqrt{\frac{288.15}{T_{t24}}} \tag{2.15}$$

The compressor outlet total temperature $T_{t3}$ is:

$$T_{t3} = T_{t24}\left[\left(\pi_c^{\frac{\kappa-1}{\kappa}} - 1\right)/\eta_c + 1\right] \tag{2.16}$$

The compressor outlet flow $W_{a3}$ is:

$$W_{a3}=W_{a24} \tag{2.17}$$

The compressor power $N_c$ is obtained:

$$N_c = W_{a24} \times c_p(T_{t3} - T_{t24}) \tag{2.18}$$

***Burner and Fuel Flow*** Now the highly pressurized air flow through high-pressure compressors enters into the burner, where it combusts with fuel spray, forming high pressure and temperature gas. This is obviously a very complicated thermal process. To simplify the analysis, it is usually disregarding the time delay effect as well as the dynamics of the injector, and simply taking the energy balance to calculate the burner outlet temperature. As the compressor outlet is assumed to be the burner inlet, the burner outlet thermal parameters can be represented as:

Burner outlet flow $W_{a4}$:

$$W_{a4} = W_{a3} + W_f \approx W_{a3} \tag{2.19}$$

Burner outlet total pressure $P_{t4}$:

$$P_{t4} = P_{t3}\sigma_b \tag{2.20}$$

where: $\sigma_b$ is the burner total pressure recovery coefficient.

Burner outlet total temperature $T_{t4}$ is approximated as:

$$T_{t4} \approx T_{t3} + \frac{Hu \cdot \eta_b \cdot W_f}{Wa_4 \cdot c_p'} \tag{2.21}$$

where: $Hu$ is the fuel oil low calorific value, $\eta_b$ is the combustion efficiency, $c'_p$ is the specific heat of gas.

***High-Pressure Turbine*** The high pressure and temperature gas coming from the burner outlet spays into the first stage guide vanes before entering high-pressure turbines. The gas expands and does work within high-pressure turbines, which consequently rotates the high-pressure compressors while providing power for engine accessories. To calculate the high-pressure turbine parameters, it is similar with compressors but in the reverse process. This is because high-pressure turbines are rigidly connected with high-pressure compressors, as low-pressure turbines are rigidly connected with the fan. Therefore, the high-pressure turbine corrected speed is defined similarly as:

$$n_{H,cor} = (n_H/\sqrt{T_{t4}})/(n_{Hd}/\sqrt{T_{t4d}}) \tag{2.22}$$

where: $T_{t4}$ is high-pressure turbine inlet total temperature, $n_{Hd}$ is high-pressure shaft speed at the design point, $T_{t4d}$ is the high-pressure turbine inlet total temperature at the design point. Then corrected flow $W_{aTH,cor}$ and efficiency $\eta_{TH}$ can be obtained from the high-pressure turbine characteristic map in terms of high-pressure turbine corrected speed $n_{H,cor}$ and high-pressure turbine pressure ratio $\pi_{TH}$.

High-pressure turbine inlet flow $W_{a4}$ is then obtained as:

$$W_{a4} = W_{aTH,cor}(P_{t4}/\sqrt{T_{t4}})/(P_{t4,ref}/\sqrt{T_{t4,ref}}) \tag{2.23}$$

where: $T_{t4,ref}$ and $P_{t4,ref}$ are the total temperature and total pressure at a reference section of the turbine inlet.

High-pressure turbine outlet total temperature $T_{t42}$ is then calculated to be:

$$T_{t42} = T_{t4}\left[1 - \left(1 - \pi_{TH}^{\frac{1-\kappa'}{\kappa'}}\right)\eta_{TH}\right] \tag{2.24}$$

High-pressure turbine outlet total flow $W_{a42}$ is:

$$W_{a42}=W_{a4} \tag{2.25}$$

High-pressure turbine power $N_{TH}$ is then:

$$N_{TH} = W_{a4} \times c'_p(T_{t4} - T_{t42}) \tag{2.26}$$

**Remark 1** It should be noted that different stages of guided vanes and rotors within the high-pressure turbines receive different amount of cooling air, and the cooling air is mixed with gas which can have significant impact on gas temperature and the power of high-pressure turbine. Meanwhile as a hot section component, the heat conduction between high-pressure turbines and gas can also influence the turbine outlet temperature.

**Remark 2** Even for the heat conduction between high-pressure turbines and gas, the heat conduction occurs among the components of blades, rotators and casings, thus heat distribution should also be considered for calculation and simulation of high-pressure turbine outlet temperature. Also, components temperature change can influence the tip clearance, which further causes the change of turbine efficiency. This should be considered as well. And this is also why active tip clearance control system is installed for almost all the modern advanced turbofan engines.

***Low-Pressure Turbine*** Gas produced in burner passes through the high-pressure turbine before entering low-pressure turbine. Within low-pressure turbines, the gas further expands and does work that power the rotation of fan and low-pressure compressors (if there are any). The calculation procedure is similar with that of high-pressure turbines, yet the factors from cooling air and tip clearance should also be taken into account.

Low-pressure turbine corrected speed $n_{L,cor}$ is defined as:

$$n_{L,cor} = (n_L/\sqrt{T_{t44}})/(n_{Ld}/\sqrt{T_{t44d}}) \tag{2.27}$$

where: $T_{t44}$ is low-pressure turbine inlet total temperature, $n_{Ld}$ is low-pressure shaft speed at the design point, $T_{t44d}$ is the low-pressure turbine inlet total temperature at the design point. Then corrected flow $W_{aTL,cor}$ and efficiency $\eta_{TL}$ can be obtained from the low-pressure turbine characteristic map in terms of low-pressure turbine corrected speed $n_{L,cor}$ and high-pressure turbine pressure ratio $\pi_{TL}$.

Low-pressure turbine inlet flow $W_{a44}$ is then obtained as:

$$W_{a44} = W_{aTL,cor}(P_{t44}/\sqrt{T_{t44}})/(P_{t44,ref}/\sqrt{T_{t44,ref}}) \tag{2.28}$$

where: $T_{t44,ref}$ and $P_{t44,ref}$ are the total temperature and total pressure at a reference section of the turbine inlet (usually take turbine inlet total temperature and pressure).

Low-pressure turbine outlet total temperature $T_{t5}$ is then calculated to be:

$$T_{t5} = T_{t44}\left[1 - \left(1 - \pi_{TL}^{\frac{1-\kappa'}{\kappa'}}\right)\eta_{TL}\right] \tag{2.29}$$

Low-pressure turbine outlet total flow $W_{a5}$ is:

$$W_{a5} = W_{a44} = K'_m P_{t6} A_6 q(Ma_6)/\sqrt{T_{t6}} = W_{a6} \tag{2.30}$$

where: $A_6$ is the nozzle outlet area for inner bypass; $q(Ma_6)$ is the flow volume coefficient, and $q(Ma_6) = Ma_6\left[\frac{2}{\kappa'+1}\left(1+\frac{\kappa'-1}{2}\right)Ma_6^2\right]^{-\frac{\kappa'+1}{2(\kappa'-1)}}$, $K'_m = \sqrt{\frac{\kappa'}{R}\left(\frac{2}{\kappa'+1}\right)^{\frac{\kappa'+1}{\kappa'-1}}}$, $T_{t6} = T_{t5}$, $P_{t6} = P_{t5} \cdot \sigma_{it}$, and $\sigma_{it}$ is the inner bypass pressure recovery coefficient.

Then the inner bypass nozzle outlet static pressure $P_{s5}$ is calculated to be:

$$P_{s5} = P_{t5} / \left(1 + \frac{1-\kappa'}{2} Ma_6^2\right)^{\frac{\kappa'}{\kappa'-1}} \tag{2.31}$$

***Bypass*** For turbofan engines, the air flow passing through the fan blades is split into two parts, one section goes into the compressors while the other section goes through the outer bypass (usually simply denoted as bypass, and the outer bypass compares with inner bypass meaning the section of air flowing through compressors, burner and turbines).

Bypass outlet total pressure $P_{t16}$ is calculated to be:

$$P_{t16} = P_{t13} \cdot \sigma_{bp} \tag{2.32}$$

where: $\sigma_{bp}$ is the bypass total pressure recovery coefficient.

Bypass outlet total temperature $T_{t16}$ is assumed to be the same with fan outlet total temperature $T_{t22}$:

$$T_{t16} = T_{t13} = T_{t22} \tag{2.33}$$

Bypass outlet flow $W_{a16}$ is:

$$W_{a16} = K_m \frac{P_{t16} A_{16} q(Ma_{16})}{\sqrt{T_{t16}}} = W_{a13} \tag{2.34}$$

where: $Ma_{16}$ is determined by $P_{s16} = P_{t16} / \left(1 + \frac{\kappa-1}{2} Ma_{16}^2\right)^{\frac{\kappa}{\kappa-1}} = P_{s6}$, and $A_{16}$ is the bypass outlet area, and $K_m = \sqrt{\frac{\kappa}{R}\left(\frac{2}{\kappa+1}\right)^{\frac{\kappa+1}{\kappa-1}}}$.

***Mixer*** Gas flow from the turbines mixes with that from bypass in mixer. Therefore:

Mixer outlet flow $W_{a7}$ can be calculated:

$$W_{a7} = W_{a6} + W_{a16} \tag{2.35}$$

Mixer outlet total pressure $P_{t7}$ is:

$$P_{t7} = \sigma_m (P_{t16} W_{a16} + P_{t5} W_{a5}) / W_{a7} \tag{2.36}$$

where: $\sigma_m$ is the mixer total pressure recovery coefficient.

Mixer outlet total temperature $T_{t7}$ can also be computed:

$$T_{t7} = (T_{t16}c_p W_{a16} + T_{t5}c'_p W_{a5})/(W_{a7}c'_p) \tag{2.37}$$

***Nozzle*** Now the static pressure of the gas from the mixer is still lower than the static air pressure, and the main function of the nozzle is thus making the gas expand and work, increasing gas static pressure, decreasing gas flow speed, and transforming the gas potential work into mechanical work for engine propulsion. Meanwhile, the high temperature gas decreases significantly its temperature passing through the nozzle, increasing engines stealth capability. The corresponding parameters are then obtained as follows.

Nozzle total temperature $T_{t8}$:

$$T_{t8} = T_{t7} \tag{2.38}$$

Nozzle total pressure $P_{t8}$:

$$P_{t8} = P_{t7}\sigma_{nz} \tag{2.39}$$

where: $\sigma_{nz}$ is the nozzle total pressure recovery coefficient.

Nozzle outlet flow $W_{a8}$:

$$W_{a8} = W_{a7} \tag{2.40}$$

Nozzle practical pressure ratio $\pi_{nz}$:

$$\pi_{nz} = P_{t8}/P_{s8} \tag{2.41}$$

Nozzle pressure ratio $\pi_{nz,us}$:

$$\pi_{nz,us} = P_{t8}/P_{s0} \tag{2.42}$$

Nozzle critical pressure ratio $\pi_{cr}$:

$$\pi_{cr} = [(\kappa + 1)/2]^{\frac{\kappa}{\kappa-1}} \tag{2.43}$$

Nozzle outlet static pressure $P_{s8}$, flow speed $v_8$, outlet Mach number $Ma_8$ should be discussed with the following two conditions:

(1) If $\pi_{nz,us} < \pi_{cr}$, nozzle works at sub-critical state:

$$P_{s8} = P_{s0} \tag{2.44}$$

$$v_8 = \varphi_{nz}\sqrt{2c'_p T_{t8}[1 - \pi_{nz} - (\kappa' - 1)/\kappa']} \tag{2.45}$$

$$Ma_8 = \sqrt{\frac{2}{\kappa' - 1}\left(\pi_{nz}\frac{\kappa' - 1}{\kappa'} - 1\right)} \tag{2.46}$$

(2) If $\pi_{nz,us} \geq \pi_{cr}$, nozzle works at critical or super-critical state:

$$P_{s8} = P_{t8}/\pi_{cr} \tag{2.47}$$

$$v_8 = \varphi_{nz}\sqrt{2c'_p T_{t8}[1 - \pi_{nz} - (\kappa' - 1)/\kappa']} \tag{2.48}$$

$$Ma_8 = 1.0 \tag{2.49}$$

where: $\varphi_{nz}$ is the nozzle speed coefficient.

The above are the components properties. The engine dynamics can then be obtained by "assembling" components together and the assembly is defined by a series of equations producing engine dynamics.

### 2.1.2 Aeroengine Working Equations

Along the gas path, the components are connected and hence the parameters are transmitted as defined in the above equations. There are then equations governing the coordinated working among these components to produce thrust. These are working equations that can be classified as algebraic working equations such as continuum of flow between two components, equalization of pressure across the same engine section, and dynamics equation such as the power balance equations for components rigidly connected with the same shaft. These working equations are often classified into steady state equations and dynamic equations, referring directly and conveniently to the engine working state.

(1) **Engine steady state equations**: when engine works at steady state, mass flow continuum and static pressure balance should be satisfied across each section of the engine, together with the power balance for rotating components.

***The power balance equation*** for fan and low-pressure turbines:

$$e_1 = (\eta_{TL} \cdot N_{TL} - N_f)/N_f \tag{2.50}$$

where: $N_{TL}$ is the low-pressure turbine power, and $N_f$ is the fan power, with $\eta_{TL}$ denoting low-pressure turbine mechanical efficiency; $e_1$ is the power residual that should approach zero.

***The power balance equation*** for compressors and high-pressure turbines is similarly expressed as:

$$e_2 = (\eta_{TH} \cdot N_{TH} - N_{ext} - N_c)/N_c \tag{2.51}$$

where: $N_{TH}$ is the high-pressure turbine power, $N_c$ is the high-pressure compressor power, with $\eta_{TH}$ denoting high-pressure turbine mechanical efficiency; $N_{ext}$ represents the high-pressure turbine shaft axial power; $e_2$ is the power residual that should approach zero.

***Static pressure balance*** at the outlet of inner and outer bypass:

$$e_3 = (P_{s6} - P_{s16})/P_{s16} \tag{2.52}$$

where: $p_{s6}$ is the inner bypass outlet static pressure, $p_{s16}$ is the outer bypass outlet static pressure. $e_3$ is the pressure residual that should approach zero.

***Nozzle flow continuum equation***

$$e_4 = (W_{a8c} - W_{a8})/W_{a8} \tag{2.53}$$

where: $W_{a8}$ is the nozzle throat flow, $W_{a8c}$ is the nozzle throat flow that is computed theoretically. $e_4$ is the flow residual that should approach zero. Other continuum equations include:

***High-pressure turbine inlet flow continuum equation***

$$e_5 = (W_{a4c} - W_{a4})/W_{a4} \tag{2.54}$$

where: $W_{a4c}$ is the corrected flow as computed through high-pressure turbine flow characteristic table, while $W_{a4}$ is the flow actually entering high-pressure turbines. $e_5$ is the flow residual that should approach zero.

***Low-pressure turbine inlet flow continuum equation***

$$e_6 = (W_{a44c} - W_{a44})/W_{a44} \tag{2.55}$$

where: $W_{a44c}$ is the corrected flow as computed through low-pressure turbine flow characteristic table, while $W_{a44}$ is the flow actually entering low-pressure turbines. $e_6$ is the flow residual that should approach zero.

(2) **Engine dynamic equations**: when engine works over a transient process, e.g. acceleration or deceleration, the flow continuum equations should still be satisfied, however the power balance equations need being modified to account for the dynamic processes. Thus, for engine dynamic processes, Eqs. (2.52), (2.53), (2.54) and (2.55) should still hold, while the shaft dynamics are augmented as the working equations.

***High-pressure shaft dynamical equation***

$$\dot{n}_H = (\eta_{TH} N_{TH} - N_{ext} - N_C) / \left[ n_H J_H \left( \frac{\pi}{30} \right)^2 \right] \tag{2.56}$$

where: $J_H$ is the rotational inertia for high-pressure shaft, $N_{TH}$, $N_C$, $\eta_{TH}$, and $N_{ext}$ are the corresponding parameters as defined in (2.51).

***Low-pressure shaft dynamical equation***

$$\dot{n}_L = (\eta_{TL} N_{TL} - N_f) / \left[ n_L J_L \left( \frac{\pi}{30} \right)^2 \right] \tag{2.57}$$

where: $J_L$ is the rotational inertia for low-pressure shaft, while $N_{TL}$, $N_f$, and $\eta_{TL}$ are defined as in (2.50).

### *2.1.3 Solving Engine Working Equations*

Engine states are obtained by solving the engine working equations simultaneously. Usually, initial conditions are provided for determination of the operation point. These conditions are called initial guess values, and they have direct influence on the performance of solutions in terms of convergence speed. Indeed, an inappropriate guess can even lead to divergence of the solution to engine working equations, albeit often result in large number of iterations for convergence.

For steady state solutions, the six parameters as the low-pressure rotational speed $n_L$, high-pressure rotational speed $n_H$, fan pressure ratio coefficient $\pi_f$, compressor pressure ratio coefficient $\pi_c$, low-pressure turbine pressure ratio $\pi_{TL}$ and high-pressure turbine pressure ratio $\pi_{TH}$ are chosen as initial guess values; otherwise the four parameters $\pi_f$, $\pi_c$, $\pi_{TL}$, and $\pi_{TH}$ are chosen as initial guess values for calculation of engine transient or dynamic states.

To proceed, for steady state, solution to the six steady state working equations leads to the solution to the errors $|e_i| \leq 10^{-6} (i = 1, 2, 3, 4, 5, 6)$. The six error equations are nonlinear and no analytical solutions can be found. Henceforth numerical solutions have to be sought, e.g. Newton–Raphson method. The Newton–Raphson method solves the engine working equations iteratively along the gradient direction, until the error equations are satisfied within a pre-determined confidence level. The outputs are the steady states of engine model working at certain steady state operational condition.

Similarly, when engine works on a transient state over the flight envelope, the four equations on mass flow continuum, pressure balance etc. should be satisfied while the corresponding error equations $|e_i| \leq 10^{-6}(i = 3, 4, 5, 6)$ are solved. The dynamical equations for the two shaft rotational speeds are approximated for the engine acceleration or deceleration dynamics. The calculation also uses Newton–Raphson iteration method, and the outputs are the corresponding dynamical states of engine model working over a transient process.

The engine working equations represent the complicated thermal–mechanical-dynamical process of engines, and thus descript the engine inherent properties. These mathematical descriptions have strict physical and chemical meaning, and the correspondence should be established. Solving the engine working equations are at the key to aircraft engine modeling techniques.

### 2.1.4 Aeroengine Performance Equations

The above are the components properties and working equations at both steady state and dynamic state. And to carry out engine performance analysis, performance indices should be defined. However, evaluating engine performance is a complicated task as there are many indices associated with reliability, life, cost, time etc. are difficult to determine qualitatively. To simplify the discussion, engine thrust and the specific fuel consumption rate is often used as engine performance parameters. The are defined as follows:

Engine thrust $F$:

$$F = Wa(v_8 - v_0) + (P_{s8} - P_{s0})A_8 \tag{2.58}$$

Engine specific fuel consumption rate $sfc$ defined as fuel consumption to produce unit thrust:

$$sfc = W_f/F \tag{2.59}$$

techniques.

### 2.1.5 Aeroengine Performance Simulations

The above sections have explained the engine properties and working equations, together with performance indices as engine thrust $F$ and specific fuel consumption rate $sfc$. Now given the corresponding coefficients in the equations and specify the characteristics maps for fan, compressor, and turbines, one can simulate the gas path sectional variables across each component and the engine performance as well.

***Steady-state Performance Simulation***

For certain type of turbofan engine, for conditions: $H = 0$ km, $Ma = 0$, and fuel flow input 0.6 kg/s, the steady performance is shown in Fig. 2.2. It is seen that the high-pressure rotational speed stabilizes after 1.5 s, and the engine reaches steady state.

Iterating with the fuel flows, and marking the corresponding steady-state performance variables on the chart, one can obtain the steady-state performance for the turbofan engine, as illustrated in Fig. 2.3. It is seen from (a) and (b) that both the high-pressure rotational speed and the low-pressure rotational speed at steady-state increase with more fuel injections; a similar trend exists for thrust, which explains that engine thrust can be "inferred" from rotational speeds; finally, (d) shows that SFC performance, which is a typical "inverted bell curve" and the SFC first decreases and then rises with an increasing fuel flow.

***Dynamic Simulations***

Aeroengines are required to frequent switching among different operational conditions over the flight envelope, e.g. from idle to takeoff, cruise to idle etc. More switching is needed for military engine maneuvers. When engine is in transient operation, dynamic models must be considered for performance analysis. Thus, dynamic

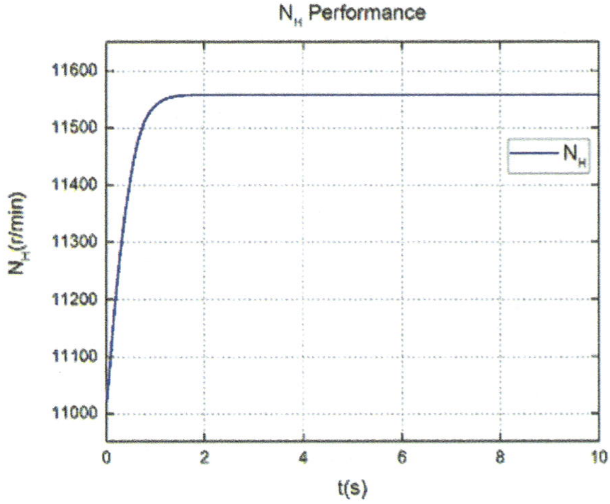


**Fig. 2.2** High-pressure rotational speed for a turbofan engine

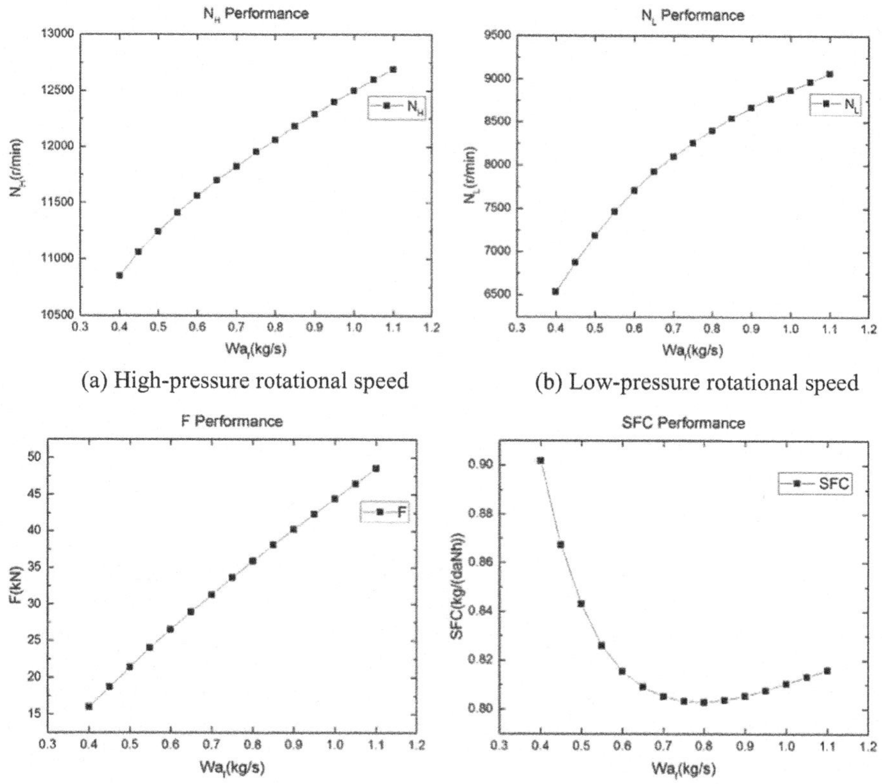


(a) High-pressure rotational speed

(b) Low-pressure rotational speed

(c) Thrust

(d) Specific fuel consumption

**Fig. 2.3** Turbofan engine steady-state performance

simulations refer to acceleration and deceleration processes, e.g. increasing fuel injection into the burner leads to acceleration while decreasing fuel injection to deceleration. During deceleration, the work load decreases, and it seems that the engine is in safe operation. This is not always true since a quick decrease in fuel injection can cause the dangerous situation of flameout, due to the unmatched fuel to air ratio in comparison to the still-high spool speed. There also exists the case for "deceleration surge" which is not usually noticed by the designers but this deceleration surge must be prevented from engine entering yet another dangerous situation.

For the above referred turbofan engine, for conditions: $H = 0$ km, $Ma = 0$, and fuel flow input as shown in Fig. 2.4: fuel flow 0.5 kg/s over $t = 0 \sim 4$ s; then a linear increase from 0.5 to 1 kg/s while keeping steady for 4 s; starting from $t = 12$ s, the engine experiences a deceleration over the next 4 s with fuel flow from 1 kg/s decreasing to 0.5 kg/s. The performance is illustrated in Fig. 2.5.

In Fig. 2.5a shows the high-pressure rotational speed and it is seen that the speed quickly increases with fuel injection while remains steady-state for fixed fuel flow; it also decreases with the decrease of fuel injection. The similar trend is demonstrated

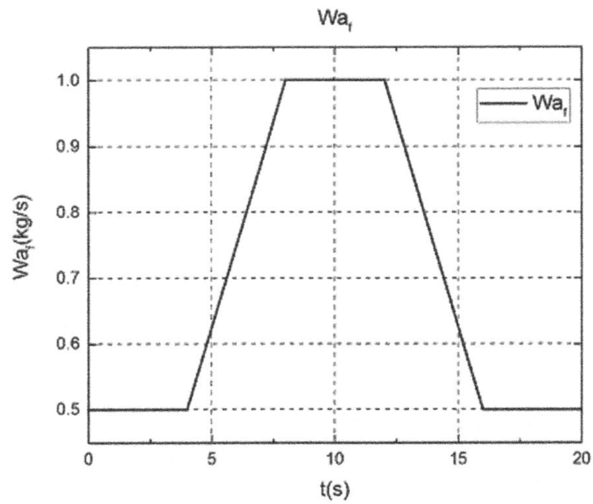


**Fig. 2.4** Turbofan engine fuel input

for low-pressure rotational speed and engine thrust in (b) and (c), respectively. The specific fuel consumption is shown in (c), and it is seen that overshoots exists during acceleration, which is typical. Specific thrust changes as thrust for a specified engine in (e). The turbine temperature is illustrated in (f), it is noted that there is an overshoot during acceleration but this is mainly due to the fact that combustion delay is ignored during the modeling process; thus, a quick increase of fuel injection into the burner leads to an over-speed of fuel-to-air ratio.

***Aeroengine Performance Analysis with Advanced Controls***

Besides these aeroengine performance analysis for both steady-state and dynamic simulations, the resulting component-level models can be further used for advanced control and fault detections, as illustrated in Fig. 2.6 for large envelope engine control, and in Fig. 2.7 for fault detection and fault tolerant control.

## 2.2 Aeroengine Linear Identification Models

This section provides a general introduction to aeroengine identification models, while alluding to the motivation moving from linear identification models to nonlinear ones. As is well-known, two common-used forms of engine models are component-level models and those of identification-based models [3]. Component-level models have been developed in the above sections. It is seen that component-level models can account for various effects such as heat soakage, turbine cooling, time delay etc. and thus they can be very accurate [1, 4–6]. Indeed, both simulations

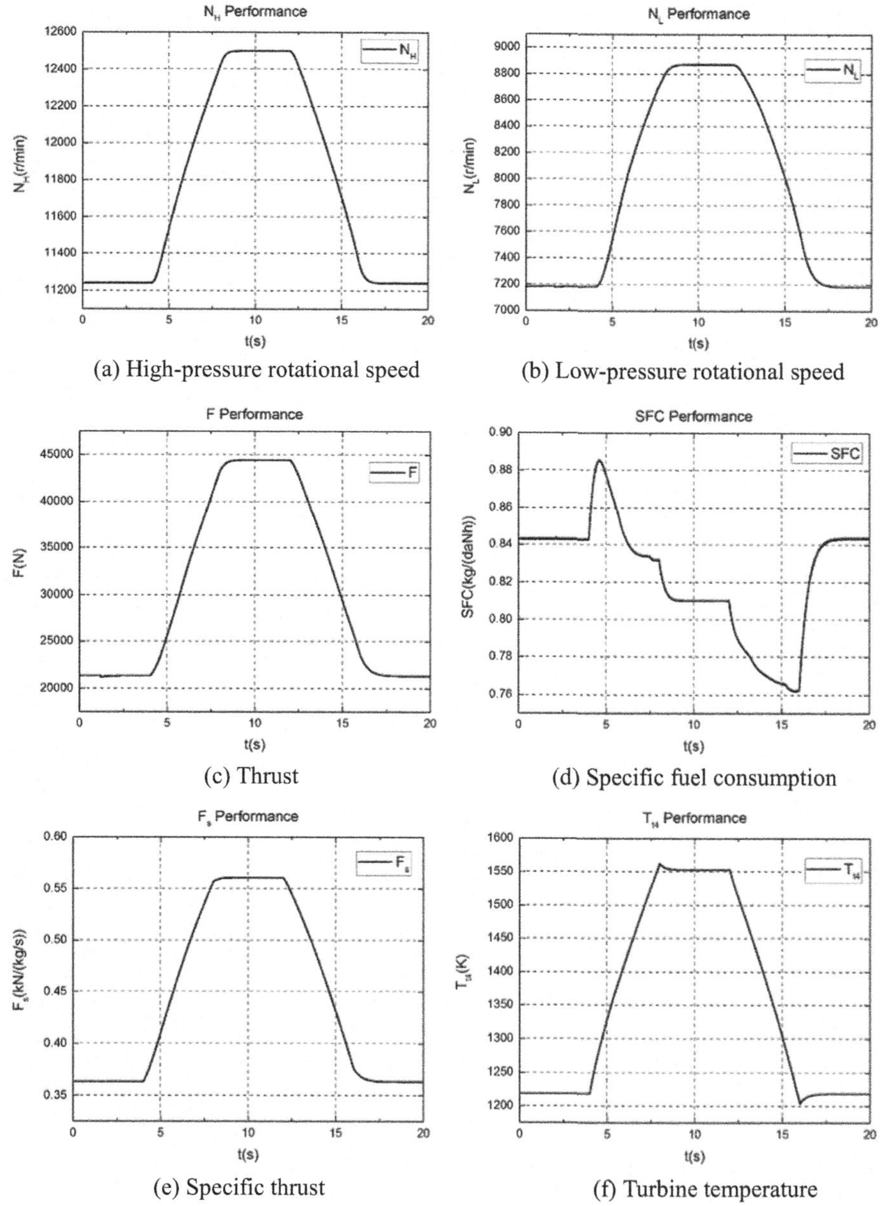


(a) High-pressure rotational speed (b) Low-pressure rotational speed

(c) Thrust (d) Specific fuel consumption

(e) Specific thrust (f) Turbine temperature

**Fig. 2.5** Turbofan engine dynamic performance

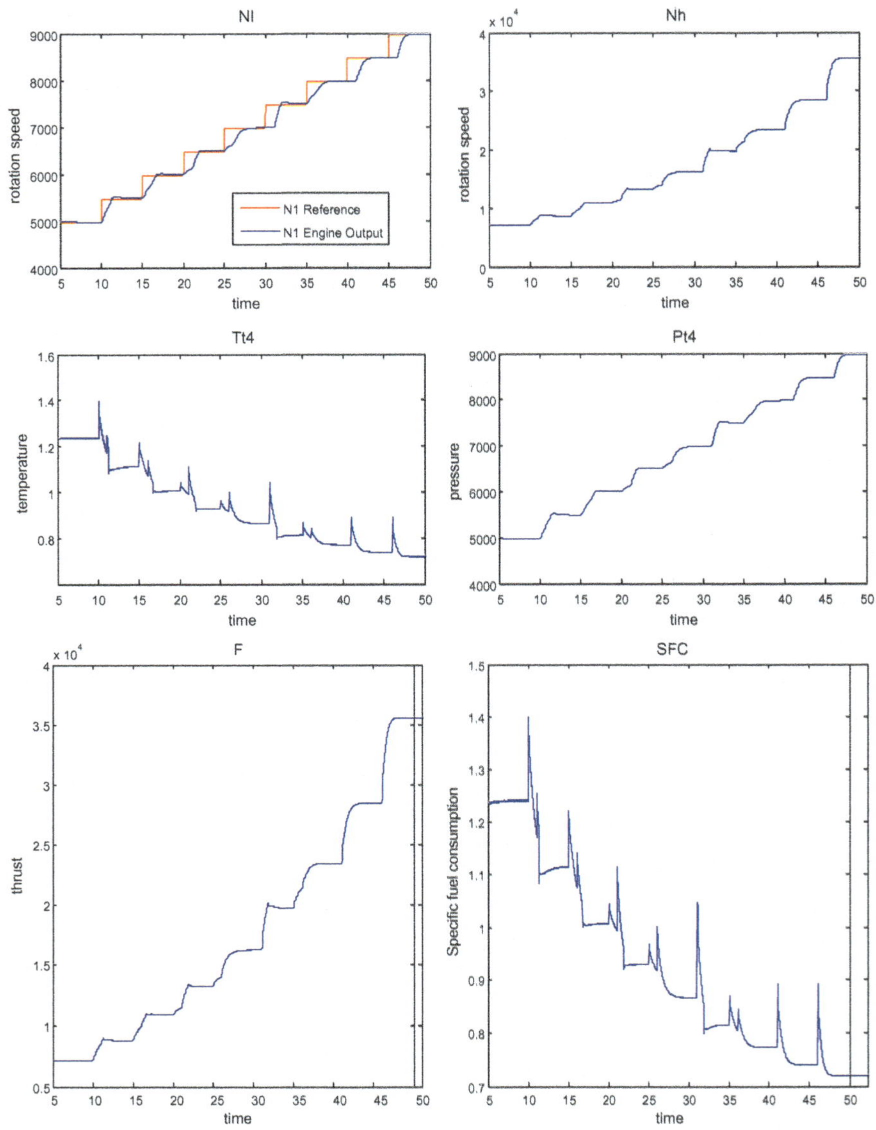


**Fig. 2.6** Large envelope engine control with component-level model

and some experimentations show remarkable accuracy with toleration in temperature, pressure and velocity within 1%. However, the accuracy sacrifices with extensive computational power due to iterative solvers [7, 8]. Therefore component-level models are more suitable for performance cycle analysis and other non-real-time applications. For those on-board applications (either control or health management), identification-based models have to be adopted due to the limited computational

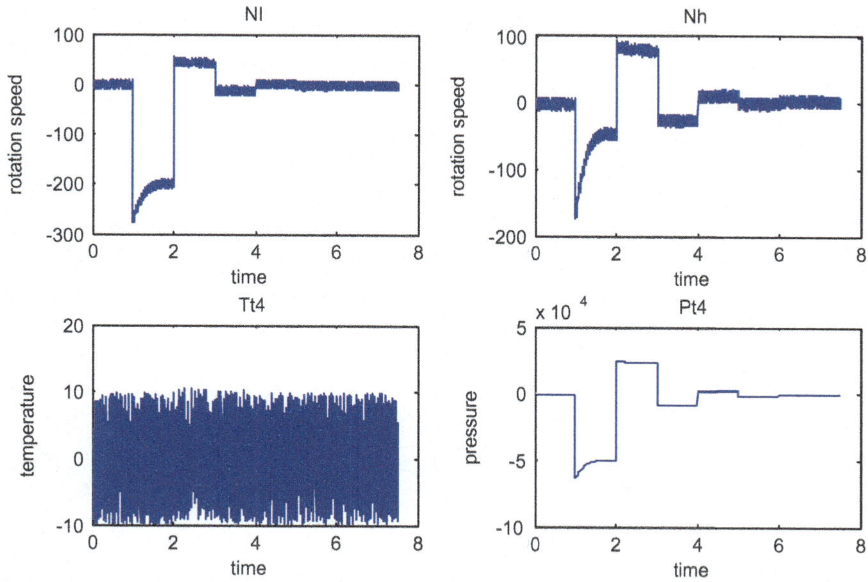


**Fig. 2.7** Component-level model for fault detection: Kalman filter

resource in engine control units (ECUs). For example, sampling rates are limited with 5–20 ms even for modern ECUs due to extreme reliability and high temperature resistance requirements.

The principle of identification-based models is to treat the engine as a "black box," and then "guess" the engine dynamics through experimental data! System identification has been a wide yet independent subject, and systematic methods have been developed, e.g. Least Squares, Generalized Least Squares, Instrumental Variables, Prediction Error Methods, Maximum Likelihood, as well as many recursive algorithms etc. As mentioned before, aeroengine design mainly uses linear models, which can be obtained through "small perturbation method"—essentially a least square method.

To see how this small perturbation method works, consider a turbofan engine model that is represented as a dynamical system:

$$\begin{cases} \dot{x} = f(x, u) \\ y = h(x, u) \end{cases} \tag{2.60}$$

where: $x(t) \in R^n$ is system state variable; $u(t) \in R^r$ is the control variable; $y(t) \in R^m$ is the output variable. Thus, when the aeroengine works around certain steady-state, e.g. around $(x_0, u_0, y_0)$, it has the following state-space model:

$$\begin{cases} \Delta \dot{x}(t) = A\Delta x(t) + B\Delta u(t) \\ \Delta y(t) = C\Delta x(t) + D\Delta u(t) \end{cases} \tag{2.61}$$

where: $\Delta \dot{x} = \dot{x} - \dot{x}_0$, $\Delta x = x - x_0$, $\Delta u = u - u_0$, $\Delta y = y - y_0 = h(x, u) - h(x_0, u)$; and $A$, $B$, $C$, $D$ are the system matrices with $x_0(t) \in R^n$, $u_0(t) \in R^r$, $y_0(t) \in R^m$ being the system variable, respectively.

Now if the state variables are chosen to be low-pressure and high-pressure rotational speed, then $x = [n_l \; n_h]^T$; control variable is usually designated to be fuel flow $u = w_{fb}$; and the output are the two rotational speeds with turbine temperature and total pressure for high-pressure turbine $y = [n_l, n_h, t_4, p_4]^T$. For this, one can obtain:

$$A = \begin{bmatrix} a_{11} & a_{12} \\ a_{21} & a_{22} \end{bmatrix}, B = \begin{bmatrix} b_1 \\ b_2 \end{bmatrix}, C = \begin{bmatrix} 1 & 0 \\ 0 & 1 \\ c_{31} & c_{32} \\ c_{41} & c_{42} \end{bmatrix}, D = \begin{bmatrix} 0 \\ 0 \\ d_3 \\ d_4 \end{bmatrix} \tag{2.62}$$

Due to the scaling of the above output variables, it is convenient to use the corrected values of $n_{hs}$, $n_{ls}$, $t_{4s}$, $p_{4s}$, $w_{fbs}$, now denoted as $N_h$ , $N_l$, $T_4$, $P_4$, $W_{fb}$ respectively with the following definition (called corrected parameters):

$$\begin{aligned} N_h &= n_h / n_{hs} \times 100\% \\ N_l &= n_l / n_{ls} \times 100\% \\ T_4 &= t_4 / t_{4s} \times 100\% \\ P_4 &= p_4 / p_{4s} \times 100\% \\ W_{fb} &= w_{fb} / w_{fbs} \times 100\% \end{aligned} \tag{2.63}$$

Now one obtains:

$$\begin{bmatrix} \Delta \dot{N}_h \\ \Delta \dot{N}_l \end{bmatrix} = \begin{bmatrix} a_{11} & a_{12} \\ a_{21} & a_{22} \end{bmatrix} \begin{bmatrix} \Delta N_h \\ \Delta N_l \end{bmatrix} + \begin{bmatrix} b_1 \\ b_2 \end{bmatrix} [\Delta W_{fb}] \tag{2.64}$$

$$\begin{bmatrix} \Delta N_h \\ \Delta N_l \\ \Delta T_4 \\ \Delta P_4 \end{bmatrix} = \begin{bmatrix} 1 & 0 \\ 0 & 1 \\ c_{31} & c_{32} \\ c_{41} & c_{42} \end{bmatrix} \begin{bmatrix} \Delta N_h \\ \Delta N_l \end{bmatrix} + \begin{bmatrix} 0 \\ 0 \\ d_3 \\ d_4 \end{bmatrix} [\Delta W_{fb}] \tag{2.65}$$

To obtain the linearized models, do the following four steps:

***Step 1***: for desired steady condition $k(N_{hk}, N_{lk}, T_{4k}, P_{4k})$, assume $A$ has eigenvalues $\lambda_1, \lambda_2$, and write:

$$\lambda_1 + \lambda_2 = a_{11} + a_{22} \tag{2.66}$$

$$\lambda_1 \cdot \lambda_2 = a_{11} \cdot a_{22} - a_{12} \cdot a_{21} \tag{2.67}$$

with a small perturbation to fuel flow $\delta W_{fb}$, the linear model has the form:

$$\begin{aligned}
\Delta N_h(t) = \delta W_{fb}\Bigg[&\frac{a_{12}b_2 - a_{22}b_1}{\lambda_1\lambda_2} + \frac{(\lambda_1 - a_{22})b_1 + a_{12}b_2}{\lambda_1(\lambda_1 - \lambda_2)}e^{\lambda_1 t} \\
&+ \frac{(\lambda_2 - a_{22})b_1 + a_{12}b_2}{\lambda_2(\lambda_2 - \lambda_1)}e^{\lambda_2 t}\Bigg] \\
\Delta N_l(t) = \delta W_{fb}\Bigg[&\frac{a_{21}b_1 - a_{11}b_2}{\lambda_1\lambda_2} + \frac{(\lambda_1 - a_{11})b_2 + a_{21}b_1}{\lambda_1(\lambda_1 - \lambda_2)}e^{\lambda_1 t} \\
&+ \frac{(\lambda_2 - a_{11})b_2 + a_{21}b_1}{\lambda_2(\lambda_2 - \lambda_1)}e^{\lambda_2 t}\Bigg]
\end{aligned} \tag{2.68}$$

$$\begin{aligned}
\Delta T_4(t) &= c_{31}\Delta N_h(t) + c_{32}\Delta N_l(t) + \delta W_{fb} * d_3 \\
\Delta P_4(t) &= c_{41}\Delta N_h(t) + c_{42}\Delta N_l(t) + \delta W_{fb} * d_4
\end{aligned}$$

And the responses are:

$$\begin{aligned}
N_h(t) &= N_{hk} + \Delta N_h(t) \\
N_l(t) &= N_{lk} + \Delta N_l(t)
\end{aligned} \tag{2.69}$$

$$\begin{aligned}
T_4(t) &= T_{4k} + \Delta T_4(t) \\
P_4(t) &= P_{4k} + \Delta P_4(t)
\end{aligned}$$

***Step 2***: At the same condition of $k(N_{hk}, N_{lk}, T_{4k}, P_{4k})$, small perturbation to the nonlinear component-level model with $\Delta W_{fb} = \delta W_{fb}$, and the nonlinear responses are:

$$\begin{aligned}
N_{h\,non}(t_k) &= N_{hk} + \Delta N_{h\,non}(t_k) \\
N_{l\,non}(t_k) &= N_{lk} + \Delta N_{l\,non}(t_k) \\
T_{4non}(t_k) &= T_{4k} + \Delta T_{4non}(t_k) \\
P_{4non}(t_k) &= P_{4k} + \Delta P_{4non}(t_k)
\end{aligned} \tag{2.70}$$

where: $t_k = 0, T, 2T, \ldots, NT$, $T$ are the sampling rate for the nonlinear model.

***Step 3***: now the two models must have the same values:

$$\begin{aligned}
\Delta N_{h\,non}(t_k) &= \Delta N_h(t_k) \\
\Delta N_{l\,non}(t_k) &= \Delta N_{lk}(t_k)
\end{aligned}$$

$$\begin{aligned} \Delta T_{4non}(t_k) &= \Delta T_{4k}(t_k) \\ \Delta P_{4non}(t_k) &= \Delta P_{4k}(t_k) \end{aligned} \tag{2.71}$$

***Step 4***: use least squares method in the above equations, the respective parameters can be obtained, leading the corresponding linear state space models.

**Example** For certain engine, at the design point of component-level model with NH = 100%, H = 0, Ma = 0, and Wf = 1205.7 kg/h. For steady state condition, perturb Wf with 5%, calculate the responses of component-level model and state space model. The two performances are compared below in Fig. 2.8.

In fact, we can perturb Wf by 1 till 6%, the corresponding responses can be calculated leading to steady state errors (Table 2.2).

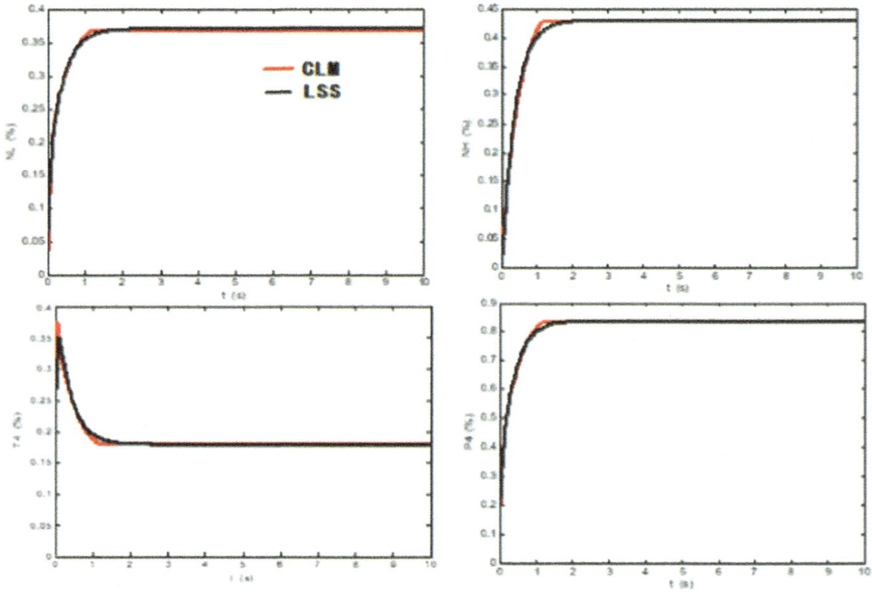


**Fig. 2.8** Performance comparison between component-level model (CLM) and linear state space model (LSS)

**Table 2.2** Model accuracy for different perturbations

| Perturb values | 1% | 2% | 3% | 4% | 5% | 6% |
|---|---|---|---|---|---|---|
| NL/% | −0.0039 | −0.0071 | −0.0076 | −0.0076 | −0.0016 | −0.0021 |
| NH/% | 0.0017 | 0.0029 | 0.0042 | 0.0045 | −0.0013 | −0.0012 |
| T4/% | 0.0027 | 0.0031 | 0.0017 | 0.0014 | 0.0023 | 0.0026 |
| P4/% | −0.0033 | −0.0028 | −0.0014 | −0.0006 | −0.0013 | −0.0018 |

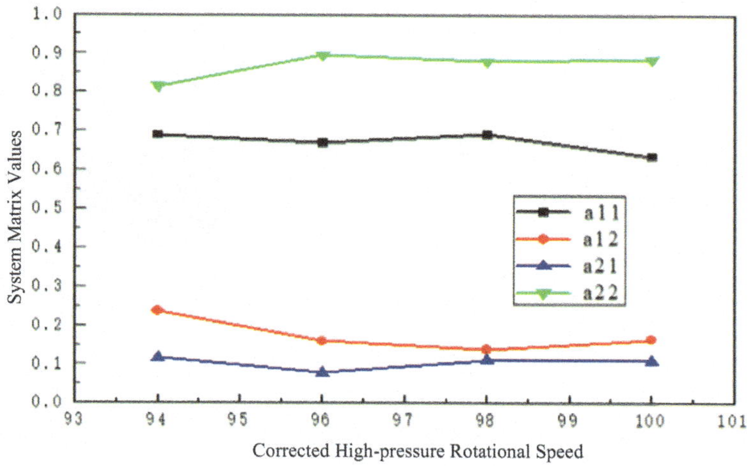


**Fig. 2.9** Linearized models can have large parameter variations

It is seen clearly that in cases from 1 to 6%, least squares method can result in acceptable fitting accuracy. However, a detailed investigation reveals that the parameters in the resulting state variables models can have relatively large variations, as shown in Fig. 2.9.

To the best of the authors' knowledge, ***there are no solutions to this large parameter variation issue***. And it is suggested to tackle this problem with the following avenues:

(1) Use transfer function models: yet it is warned that this usually leads to a model with high orders and thus it is hard to guarantee accuracy; or
(2) Use of identification-based methods, particularly nonlinear identification models.

This is one of the motivations to the following sections and to the project "model-based nonlinear control of aeroengines" in general.

## 2.3 Aeroengine Nonlinear Identification Models for Advanced Control

Indeed, component-level models are essentially not control-oriented. And it is usually not possible to synthesis a controller based on a component-level model for most of the control design methodologies. This necessitates identification-based methods to establish models for control design purposes. Besides the conventional least-squares-based small perturbation method [9, 10], a variety of identification approaches have been utilized for engine modeling, e.g. maximum likelihood method is used for engine frequency domain identification [11]; scheduling least squares method is

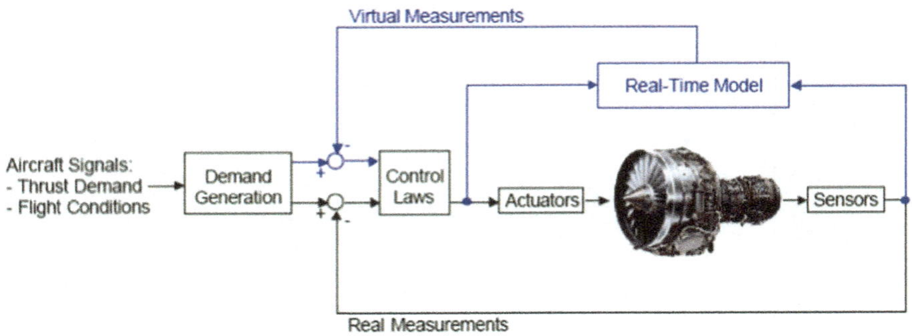


**Fig. 2.10** Virtual measurement for direct thrust control

developed for identifying linear parameter dependent models [12] and decentralized piecewise linear models [13] etc. However, most of the resulting models are essentially linear fitted for corresponding linear control design methodologies such as $H_2/H_\infty$ controls [14, 15]. Due to the inherent nonlinearity in aeroengine dynamics, identification-based nonlinear models have been developed in recent years. Nonlinear autoregressive with exogenous inputs (NARX) models have been shown to possess satisfactory prediction with experimental data during gas turbine normal operation [16]; Hammerstein-Wiener models have been demonstrated to be suitable for on-board applications [17]; novel generalized describing function models can accelerate turbine engine simulations while suitable for investigation of nonlinear control systems [18]; Neural networks and wavelet-based neural networks models have been shown to apply over a full flight envelope [19, 20]; even evolutionary algorithms based adaptive modeling techniques are developed with performance improvement [21].

In fact, identification-based nonlinear methods can obtain both "control-oriented on-board real-time models" and "fault detection-oriented models," which is at the core of advanced control and health management, e.g. thrust/power management, stability/surge margin estimation etc. An example is illustrated in Fig. 2.10. As mentioned above, nonlinear identification models have the potential to address the parameter variation problem. It is also worth mentioning that nonlinear identification models are also pertinent to onboard real-time applications. Usually, an on-board real time model consists of a series of linear models which is only feasible within limited flight envelope. Effectiveness and accuracy are two key factors for practical application. However, as commented above, due to the inherent nonlinearity of engine dynamics, it is natural to utilize nonlinear identification methods to establish on-board real time models.

In this section, however, Hammerstein-Wiener models will be presented first to demonstrate its feasibility for on-board applications; then a comparative study is to be carried out where BP neural networks (BPNN), Improved BP neural networks (IBPNN), and Piecewise NARX methods are utilized for identification of a generic two-spoon high-bypass turbofan engine models. This forms the key contribution of the current proposal where comparative studies are rarely seen in the field. While the paper particularly concerns applications to engines, the general theories upon neural networks and NARX can be consulted to many well-known references, e.g. [22, 23].

## 2.4 Aeroengine Hammerstein-Wiener Modeling

### 2.4.1 Introduction to Hammerstein-Wiener Systems

The Hammerstein-Wiener system is represented by a cascade connection of a static nonlinearity *NL1*, a linear dynamic system *LS* that is followed by another static nonlinearity block *NL2* (Fig. 2.11). The cascade connection *NL1* and *LS* is called the Hammerstein system while the cascade connection of *LS* and *NL2* is called the Wiener system. Therefore, the Hammerstein-Wiener system includes both Hammerstein and Wiener systems as special cases. Although these models are the simplest types of block-oriented nonlinear systems, they appear in many engineering applications. Indeed, the Hammerstein-Wiener model representation has long been used in process industries, biological processes and communication systems etc., due to its reported capability to modelling a large number of nonlinear systems (see [24–26] for example), and it is therefore an irresistible impetus to adopt the Hammerstein-Wiener representation for aircraft engine modelling. However, a detailed, yet careful analysis has to be carried out to justify the feasibility of modelling of aircraft engines using H-W representation. Theoretically, the identification of both Hammerstein and Wiener systems has been investigated intensively, to a lesser extent for identification of Hammerstein-Wiener systems [27, 28]. But many iterative algorithms have been proposed to identify the Hammerstein-Wiener systems and estimators exist in commercial software such as MATLAB, where the nonlinearity class can be piecewise linear, one layer sigmoid network, wavelet network, saturation, dead zone, one-dimensional polynomial, even with user-defined nonlinearities [29].

Therefore, the Hammerstein-Wiener system can be a black-box structure providing flexible parameterization of nonlinear models. This makes the Hammerstein-Wiener model find applications in modelling electro-mechanical systems and radio frequency components, audio and speed processing, chemical processes etc. The block representation and transparent relationship to linear systems are more attractive than those nonlinear models such as neural networks and Volterra models, which are relatively heavy-duty and computationally demanding. However,

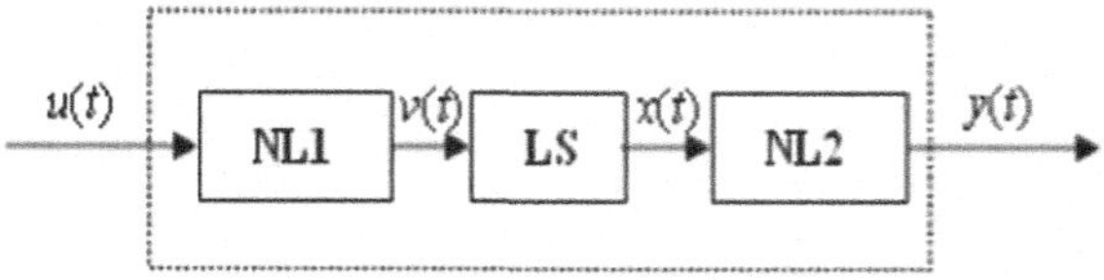


**Fig. 2.11** Hammerstein-Wiener system representation

even with these desirable properties, the application of Hammerstein-Wiener representation to the modelling of aircraft engines has to suffer a careful study which is turned to in the next subsection.

### 2.4.2 *Feasibility Analysis*

It is seen that the Hammerstein-Wiener system is represented by two static nonlinear blocks "sliced" by a linear subsystem. To model the aircraft engines, first identify $u(t)$ and $y(t)$ in Fig. 2.11 as the fuel flow ratio $w_f$ and engine shaft rotational speed $N$, respectively. That is, we are controlling the engine speed with fuel flow ratio. First start with components analysis, it is well-known that at certain (steady state) operation condition, both the compressors and turbines work at certain position along the working lines, whose typical characteristic map is shown in Fig. 2.12 for compressors and similarly for turbines. Now increasing the fuel flow to the combustor will cause an increase in the gas temperature that is proportional to the amount of fuel being injected. This linearity would conform with the engine rotational speed $N$, should the fuel flow ratio $\Delta w_f$ be small, e.g. within 1% of fuel flow change. However, once the deviation of $\Delta w_f$ is large, significant nonlinearity emerges as both the compressors and turbines will work at different position upon their characteristic maps. In fact, this nonlinearity can be represented by a series of piecewise linear functions and this justifies the widespread usage of linear gain scheduling and linear parameter varying (LPV) controls. As a consequence, viewing the engine as an

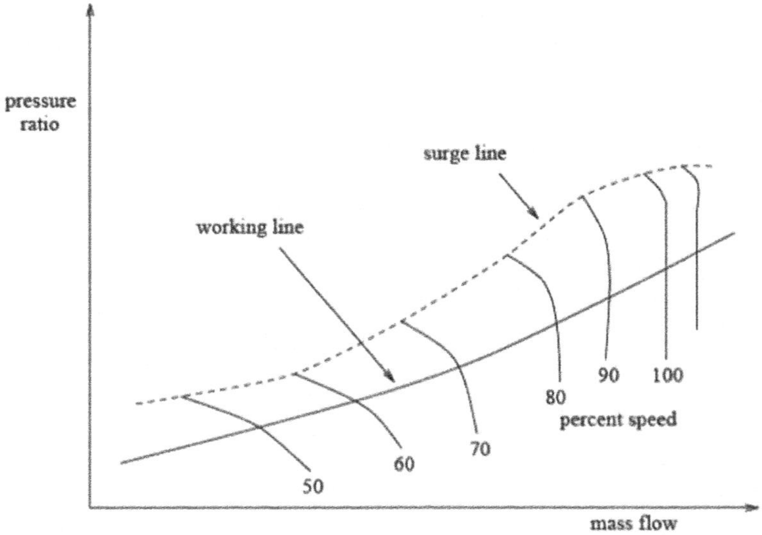


**Fig. 2.12** Overall compressor characteristics over the operational speed range. Adopted from [30]

integrated system (changing dynamics with scheduling variables H and Ma, or T1, P1), the Wiener system can be capable of representing the aircraft engine model, at least over relatively larger flight envelope than a linear model could be.

From the control system point of view, one of the key functions is limit protection, e.g. preventing the engine from surge, from flaming, from reaching mechanical limits etc. Also, the actuators have limits such as saturation. This nonlinearity can be represented by the Hammerstein system.

That is, *NL1* can usually be chosen to be saturation nonlinearities while *NL2* is designated to be piecewise linear functions. Therefore, using the Hammerstein-Wiener system can indeed represent the aircraft engine model, at least over an extended flight envelope. Such an attempt will be made in the next section.

### 2.4.3 Model Identification over a Limited Flight Envelope

It is well-known that engine controls are gain-scheduled and normally six or more controllers are usually scheduled between idle and full power [30–33]. Approximately 5% of the rotational speed of scheduling variable is covered by each controller. To demonstrate the capability of the Hammerstein-Wiener modelling we shall cover 10% variation of the rotational speed. If this can be done, it will show that a nonlinear model can extend the feasible region comparing with that a linear model, hence simplifying the scheduling logic. This is often overlooked in the literature and this note will thus contribute to this line of research.

#### *Data Preparation*

Engine performance data should be prepared before the identification and validation processes. For model identification, while the model structure and model order can be determined through an iterated process, one thing that is often overlooked is the excitation function (or the time history of input data) that is used to generate the outputs. In fact, a good excitation function should cover the positive as well as the negative side of the nominal value of an input variable. Ideally, there should be at least two crossovers in the excitation function [34]. In terms of this remark, the data are generated as shown in Fig. 2.13. The model used is a component-level model that has been built recently by the authors for a generic two-spool, turbo-fan engine [35].

#### *Model Identification and Validation*

We shall use the acceleration data from 5 to 20 s for model identification and use the deceleration data from 20 to 35 s for model validation. Now observe the input/output data as shown in Fig. 2.13, then either *NL1* or *NL2* subsystem can be designated as saturation nonlinearity; the other one is chosen to be piecewise linear functions considering the gain-scheduling nature of an aircraft engine model; the linear subsystem is usually chosen to be a three-order linear model with unity delay. Given these considerations, the key to the Hammerstein-Wiener modelling of engine

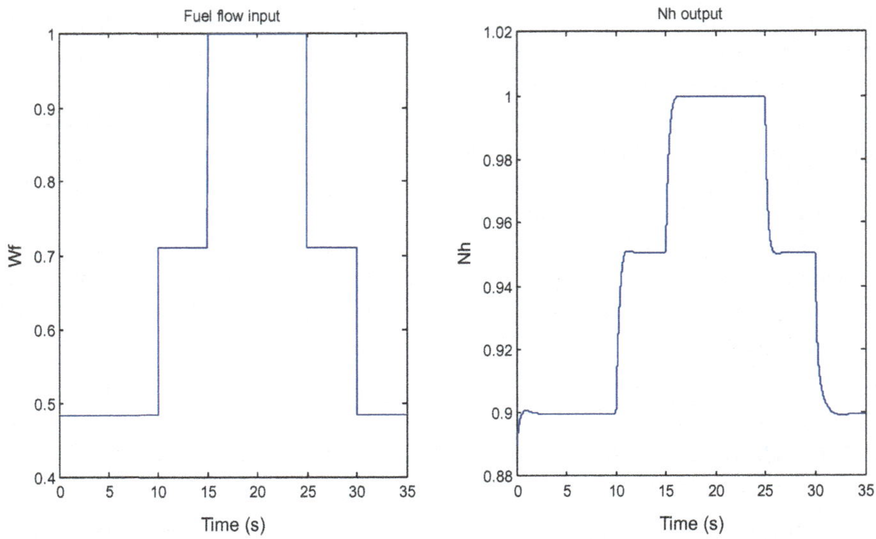


**Fig. 2.13** Data prepared for identification and validation using the component-level model in [35]. The data are obtained at sea level and static condition

is to determine the saturation position (*NL1* or *NL2*) and the order/number of the corresponding piecewise linear functions. The result for different options is shown in Fig. 2.14. It is seen that the m2 model (*NL1*: saturation nonlinearity [0.485, 1.0]; *NL2*: piecewise linear function of order 8) has an over 90% of fit with a loss function only 5.2684e−005. Therefore, it is remarkable that a single Hammerstein-Wiener model can cope with 10% variation of the rotational speed, significantly extending the feasible region of the model and consequently simplifying the switching logic comparing with those linear models that may only be valid within 3–5% of rotational speed variation. A detailed discussion is turned to in the next subsection.

***Further Discussion***

Figure 2.14 also reveals that saturation nonlinearity for *NL2* as in m5 is not feasible. This is comprehensive considering the engine control system is essentially fuel-based and as a consequence, for Hammerstein-Wiener modelling, *NL1* can always be designated to be a saturation function while *NL2* takes piecewise linear functions, reflecting the gain-scheduling nature of the engine dynamics. This design philosophy is validated by extensive simulations and a sample of results with *NL1* being saturation nonlinearity [0.485, 1.0] while varying orders of piecewise linear function for *NL2* is shown in Table 2.3. It is seen that there is no linear correlation between the fitting degree/loss function and the order of the chosen piecewise linear function for *NL2*. Therefore, the best strategy to determine the order is through trial-and-error. Here arises an interesting question: is it possible to deduce certain characteristics of the engine as reflected by the optimal order that provides best fit?

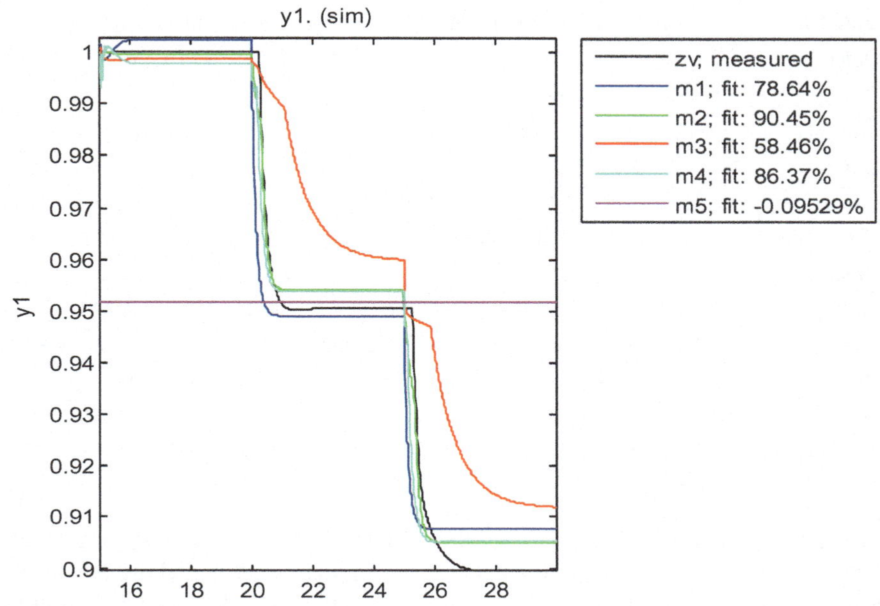


**Fig. 2.14** Difference combinations: m1–m4 have the same saturation nonlinearity [0.485, 1.0] for *NL1*, with *NL2* being piecewise linear functions with an order of 1, 8, 9, 20, respectively; m5 has a piecewise linear function of order 5 for *NL1*, with a saturation nonlinearity [0.9, 1.0] for *NL2*

**Table 2.3** Varying order of piecewise linear function for *N2* over a limited flight envelope

| Order | 1 | 2 | 3 | 4 | 5 |
|---|---|---|---|---|---|
| Fit (%) | 78.64 | 80.24 | 80.43 | 81.67 | 83.94 |
| Loss function | 7.2804e−005 | 6.4856e−005 | 6.2733e−005 | 6.3592e−005 | 6.8837e−005 |
| Order | 6 | 7 | 8 | 9 | 10 |
| Fit (%) | 84.36 | 84.75 | 90.45 | 58.46 | 82.15 |
| Loss function | 7.5602e−005 | 6.5954e−005 | 5.2684e−005 | 4.4887e−005 | 5.8726e−005 |
| Order | 11 | 12 | 13 | 14 | 15 |
| Fit (%) | 87.05 | 82.1 | 84.34 | 83.99 | 81.25 |
| Loss function | 5.9196e−005 | 5.844e−005 | 5.8273e−005 | 5.782e−005 | 6.9019e−005 |
| Order | 16 | 17 | 18 | 19 | 20 |
| Fit (%) | 86.48 | 84.86 | 87.94 | 84.89 | 86.37 |
| Loss function | 5.9525e−005 | 5.606e−005 | 5.9297e−005 | 6.3485e−005 | 5.5305e−005 |

## *2.4.4 Model Identification over Extended Flight Envelope*

To further testify the capability of the Hammerstein-Wiener technique for aircraft engine modelling, identification over an extended flight envelope is carried out. This

**Table 2.4** Varying order of piecewise linear function for $N2$ over extended flight envelope

| Order | 1 | 2 | 3 | 4 | 5 |
|---|---|---|---|---|---|
| Fit (%) | 65.64 | −11.13 | −25.43 | 32.83 | 69.15 |
| Loss function | 6.5862e−005 | 2.1605e−004 | 1.7432e−004 | 1.5001e−004 | 1.3267e−004 |
| Order | 6 | 7 | 8 | 9 | 10 |
| Fit (%) | 80.51 | 82.50 | 41.07 | 83.36 | 58.97 |
| Loss function | 1.1428e−004 | 1.1646e−004 | 4.8003e−005 | 2.9222e−005 | 2.7871e−006 |
| Order | 11 | 12 | 13 | 14 | 15 |
| Fit (%) | 85.44 | −25.46 | 85.58 | 34.90 | 84.94 |
| Loss function | 7.9356e−005 | 1.306e−006 | 8.8027e−005 | 9.1376e−007 | 5.9988e−005 |
| Order | 16 | 17 | 18 | 19 | 20 |
| Fit (%) | 44.63 | 63.70 | 34.80 | 60.12 | 37.55 |
| Loss function | 7.6104e−007 | 6.7889e−005 | 9.4572e−007 | 5.7418e−005 | 1.6919e−007 |

time, 20% speed variation of $Nh$ from 0.8 to 1.0 (80–100% TO max) is considered. Again, we still use the acceleration data for model identification and use the deceleration data for model validation. A sample of results with $NL1$ being saturation nonlinearity [0.2, 1.0] while varying orders of piecewise linear function for $NL2$ is shown in Table 2.4. Comparing with Table 2.1 for 10% variation of $Nh$ from 0.9 to 1.0 (90–100% TO max), the optimal fit drops from 90.45 to 85.58%, while the average fitting degree decreases significantly (most of the fitting degree in Table 2.1 are above 80%). Several fittings such as those with orders 9, 11, 13 and 15 can capture the engine dynamics and hence can be an acceptable model for the engine, however although not shown here their transients indeed start deterioration comparing with the model for 10% speed variation of $Nh$. Therefore, the extension of the Hammerstein-Wiener modelling technique to a large flight envelope, is promising but should meanwhile be explained carefully and be validated through more extensive simulations and possibly extreme events "training" than the brief study presented here.

**Remark** Three-layer RBF Neural Networks is recognized to be a "universal approximator", so it will also be interesting to compare Hammerstein-Wiener models with those of Neural Network-based models, e.g. in terms of accuracy and real-time performance.

### *2.4.5 Summary for Hammerstein-Wiener Aeroengine Modeling*

Hammerstein-Wiener modelling of aircraft engines has been studied in this section. It has been shown that nonlinear model can significantly extend the feasible/valid region over the flight envelope comparing with those linear models. Nonlinear models can thus have the potential to simplify the engine control logic carried onboard. For MBD and EHM, using nonlinear models to replace linear ones necessitates the study of nonlinear control techniques. This may bring some difficulty to engine control system design but its potentiality should be further evaluated. Our view is shaped by one of our previous studies [36] demonstrating the promising results of nonlinear control designs. The nonlinear techniques are initiated to reduce the switching logics in FADEC. Utilization of nonlinear design toolsets for such a purpose is in fact very important but somehow has been overlooked by the engine control community. It is hoped that this note, however brief, can call upon this line of research.

## 2.5 Aeroengine Back Propagation Neural Network Modeling

Neural network has been recognized as a universal approximator to nonlinear functions, while back propagation neural network (BPNN) is one of the widely used ones. Deployment of BPNN typically includes the following steps: (1) setup of initial weighting coefficients; (2) calculation of the outputs for given inputs; (3) calculation of performance functions; (4) determination if performance requirement is satisfied; (5) if not, carry out back propagation computation until the error falls within the predetermined confidence level.

***Data Preparation***

Based on the above training procedures, it is decided to choose the fuel flow $W_f$ and the nozzle throat area $A_8$ as inputs, while the two rotational speeds $n_H$ (high-pressure) and $n_L$ (low-pressure) as outputs. The original training data are produced by a generic two-spool, high-bypass commercial turbofan engine component-level model (CLM) that has been established beforehand. At the sea level and static condition, running of the CLM leads to the input–output data pairs shown in Fig. 2.15.

The above sample data need to be further processed: to avoid the sensitivity of parameters to modeling accuracy, the data are normalized, that is, if the maximum value is $Y_{\max}$ and for sample $Y_i$, it is to be normalized as $\widehat{Y}_i$ defined by:

$$\widehat{Y}_i = \frac{Y_i}{Y_{\max}} \times 100\% \tag{2.72}$$

After normalization, $W_f$, $A_8$, $n_H$ and $n_L$ are dimensionless and all the data will be within [0, 1]. The identified values can be anti-normalized, should they be applied for other control or monitoring purposes.

During the BPNN training process, data are randomly tri-partitioned: training data, validation data, and test data. Training data account for 75% of data set for correcting the weights and thresholds, with validation data of 15% for generating residuals and determination of the termination of iteration. The remaining test data

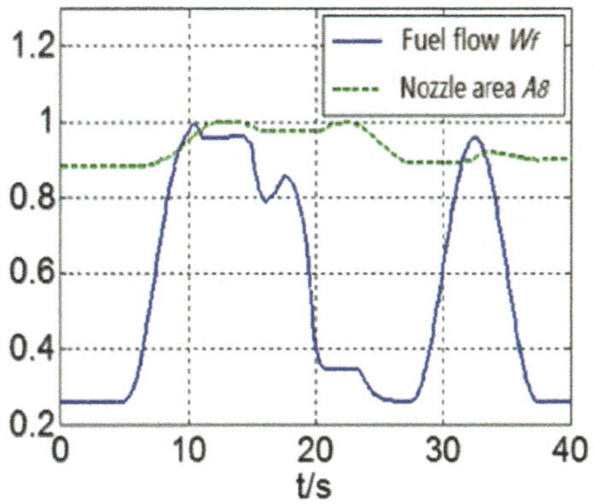


(a) Input data: $W_f$ and $A_8$

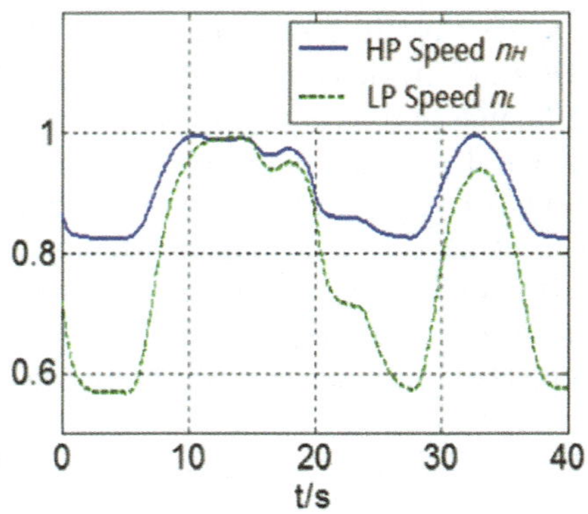


(b) Output data: $n_H$ and $n_L$

**Fig. 2.15** Input–output data pairs at sea level, static condition: the sampling rate is 0.01 s

of 15% will then be used for performance analysis and evaluation after the training process.

***Model Identification***

For BPNN with one hidden layer, the numbers of nodes in input layer, hidden layer, and output layer should be determined before the training process. Considering $W_f$ and $A_8$ as inputs while $n_H$ and $n_L$ as outputs, both input and output layers have 2 nodes; nodes in hidden layer depend on system nonlinearity and after experimentation, the number of nodes is chosen to be 10. The identification block diagram is shown Fig. 2.16.

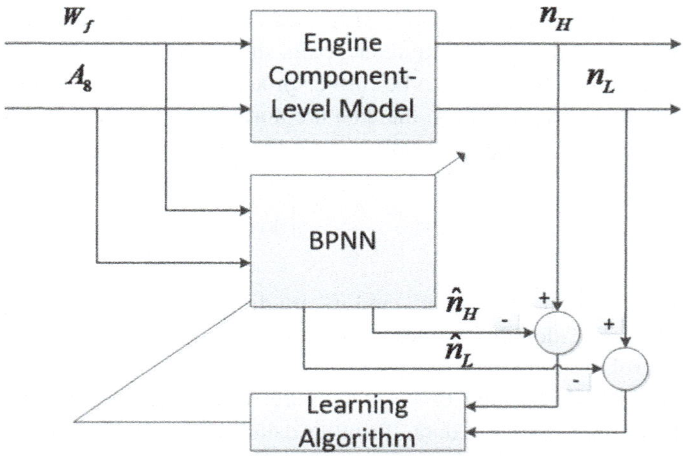


**Fig. 2.16** Aeroengine BPNN identification block diagram

***Model Validation***

After extensive training with the facility of neural network toolbox in MATLAB, the BPNN is obtained as the aeroengine model. The corresponding accuracy can be validated by varying input signals different from training data. This is shown in Fig. 2.17.

From Fig. 2.17c, d, it is seen that BPNN model identifies engine high pressure rotational speed $n_H$ very well with an accuracy within 2.3%; however, Fig. 2.17e, f shows that the error in low pressure rotational speed $n_L$ reaches up to 11%, which is usually unacceptable for dynamic performance. The result is not surprising because the engine rotational speeds are dynamic variables while BPNN identifies a static model. This necessitates the consideration of improved BPNN with dynamic consideration. This is considered in the next section.

## 2.6 Aeroengine Improved Back Propagation Neural Network Modeling

The improved BPNN (IBPNN) takes the engine dynamics into account. As turbofan engines typically have 2-order dynamics, the IBPNN structure is determined as shown in Fig. 2.18.

Thus the IBPNN has an input layer with 6 nodes: $W_f(t)$, $A_8(t)$, together with the regressors $n_H(t-1)$, $n_H(t-2)$, $n_L(t-1)$, and $n_L(t-2)$; the hidden layer is designed to have the same number of nodes with the BPNN in the last section; the output layer has two nodes $n_H(t)$ and $n_L(t)$.

With data preparation and processing as in Sect. II, the IBPNN identification model is obtained and the resulting validation result is shown in Fig. 2.19 where for space saving, only the identification error signals for high-pressure and low-pressure rotational speed are given.

It is seen from Fig. 2.19 that the IBPNN achieves the dynamic modeling accuracy within 0.2% for $n_H$ and 2% for $n_L$, which is a significant improvement comparing with BPNN. The comparison is summarized in Table 2.5.

Usually, the IBPNN model is accurate enough for initial design for advanced control and engine health management purposes. However, the IBPNN model is not control-oriented in the sense that an optimal controller is not to be synthesized based on a neural network representation. This necessitates the investigation into other identification method whose resulting model can be utilized for controller design purpose. In the next section, a piecewise NARX (nonlinear auto-regressive with extra inputs) model is to be introduced.

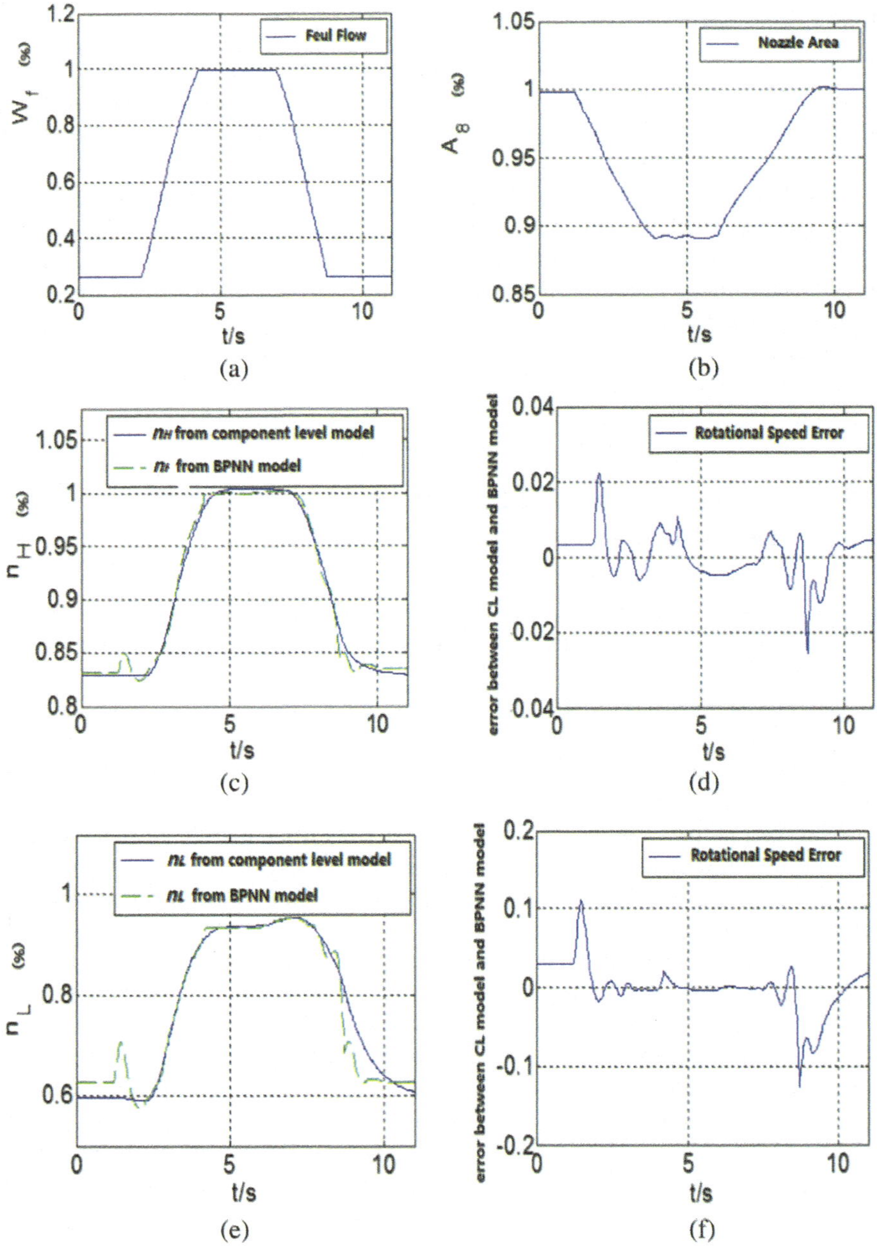


**Fig. 2.17** Model validation for BPNN—**a**, **b** are the input signals for $W_f$ and $A_8$ for validating the dynamic performance of BPNN model; **c** compares $n_H$ for component-level model output and BPNN output, with **d** denotes the error signal; similarly, **e** compares $n_L$ for component-level model output and BPNN output, with **f** denotes the error signal

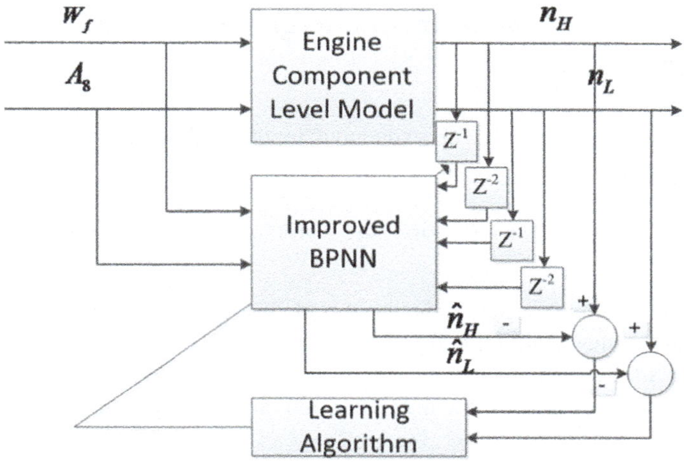


**Fig. 2.18** Aeroengine IBPNN identification block diagram

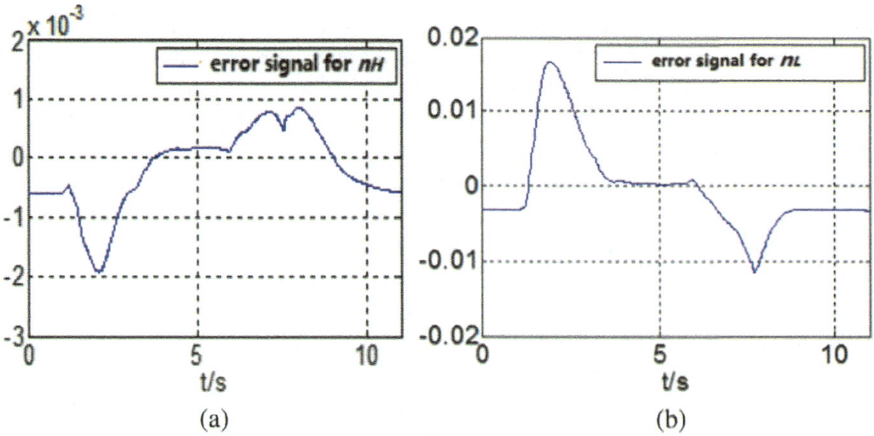


**Fig. 2.19** Model validation for IBPNN— for same input signals of $W_f$ and $A_8$ for validating the dynamic performance of BPNN model, **a** denotes the error signal between component-level output and IBPNN for $n_H$; and **b** is the error signal between component-level output and IBPNN output for $n_L$

**Table 2.5** Modeling accuracy comparison

| Model | | |
|---|---|---|
| Signal | BPNN (%) | IBPNN (%) |
| $n_H$ error | 2.3 | 0.2 |
| $n_L$ error | 11 | 2 |

## 2.7 Aeroengine Nonlinear Auto-Regressive with Exogenous Input Modeling

A nonlinear auto-regressive with exogenous input (NARX) model has the general analytical expression as:

$$y(t) = f[y(t-1), \ldots, y(t-n_a), u(t-n_k), \ldots, u(t-n_k-n_b+1)] \quad (2.73)$$

where: $y(t)$ is the output, $u(t)$ is the input, and $f$ is a nonlinear function or operator that depends on the information from previous inputs and outputs; $n_a$ and $n_b$ are the model orders for output and input respectively with $n_k$ denoting system delay. Thus, the NARX model consists of the regressors and the nonlinear estimator as shown in Fig. 2.20.

***Model Structure and Data Preparation***

With the same setup for identifying both BPNN and IBPNN models, the fuel flow $W_f$ and the nozzle throat area $A_8$ are chosen as inputs, while the two rotational speeds $n_H$ (high-pressure) and $n_L$ (low-pressure) as outputs. However, initial experiment with model identification shows that a single NARX model cannot possess enough accuracy and it is thus decided to identify a piecewise NARX (PNARX) model where two NARX representations are piecewise-connected with NARX-1 from engine idle state ($n_{H,idle} = 82.69\%$) to $n_H = 90\%$ and NARX-2 from $n_H = 90\%$ to TO ($n_H = 100\%$). Consequently, two sets of input/output data should be prepared. At the sea level and static condition, running of the CLM leads to the input–output data pairs shown in Fig. 2.21.

Again, to avoid the sensitivity of parameters to modeling accuracy, the data have been normalized as defined in Eq. (2.72). Then in Fig. 2.21, (a) and (b) input–output data are used for identifying first piece of the model named NARX-1; (c) and (d) input–output data are for the second piece named NARX-2; the final PNARX model is the two sub-models scheduled by rotational speed.

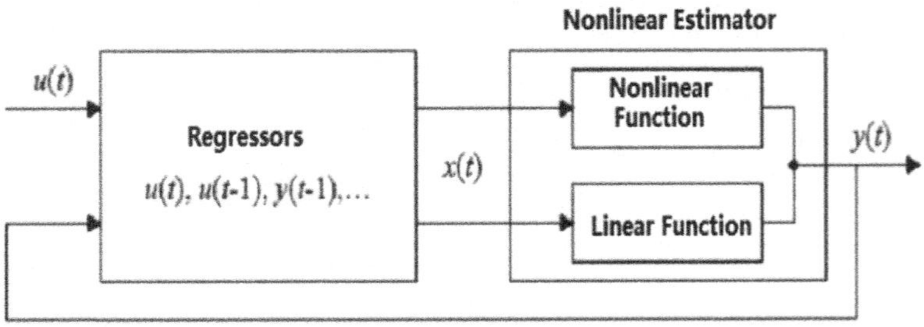


**Fig. 2.20** Typical structure of NARX model

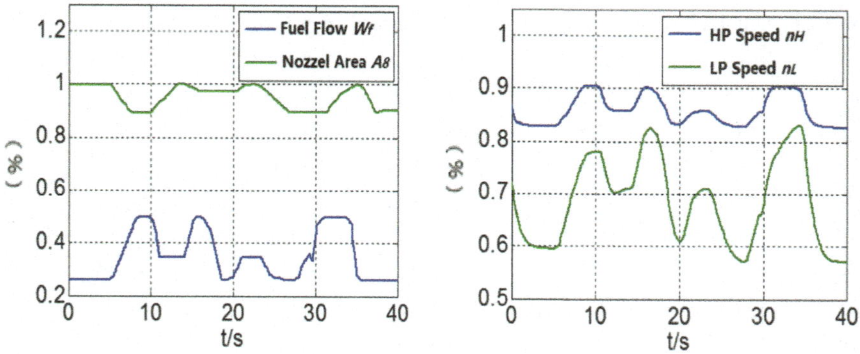


(a) Input data: $W_f$ and $A_8$ (82.69% $n_{H\ \max}$~90% $n_{H\ \max}$) (b) Output data: $n_H$ and $n_L$ (82.69% $n_{H\ \max}$~90% $n_{H\ \max}$)

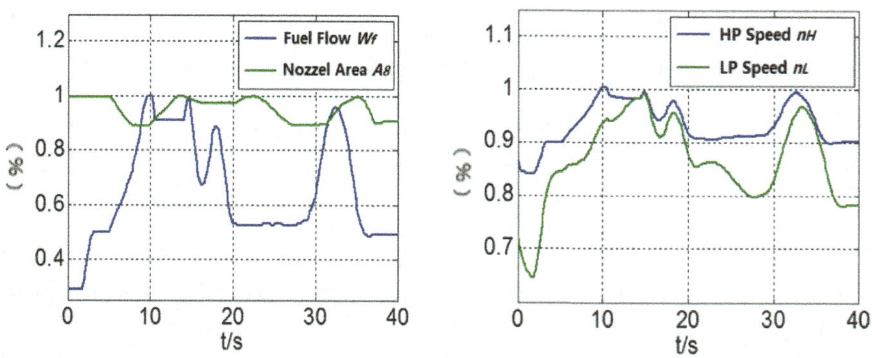


(c) Input data: $W_f$ and $A_8$ (90% $n_{H\ \max}$~100% $n_{H\ \max}$) (d) Output data: $n_H$ and $n_L$ (90% $n_{H\ \max}$~100% $n_{H\ \max}$)

**Fig. 2.21** Input–output data pairs at sea level, static condition

### *Model Identification*

Divide the above input–output data into 2 groups, 0–30 s as model identification while 30–40 s for model validation. However, for both NARX-1 and NARX-2, the corresponding parameters should be carefully tuned and extensive simulations give the following result shown in Table 2.6.

It is seen from Table 2.6 that different choices in orders ($n_a$ and $n_b$) and time delays ($n_k$) produce different models with varying degree of accuracy, where the model *m1* shows a very high accuracy in both $n_H$ and $n_L$ with best fitting 97.14% and 95.85%, respectively. Hence *m1* is chosen as the model for NARX-1. In specific:

$$n_H(t) = f[n_H(t-1), W_f(t-1), A_8(t-2)] \tag{2.74}$$

**Table 2.6** Different parameters for NARX-1 Model (82.69% $n_{H\max}$ ~ 90% $n_{H\max}$)

| No | $n_a$ | $n_b$ | $n_k$ | $n_H$(Fit %) | $n_L$(Fit %) |
|---|---|---|---|---|---|
| *m1* | [1 0;0 1] | [1 1;1 1] | [1 2;1 2] | 97.14 | 95.85 |
| *m2* | [1 0;0 1] | [2 2;2 2] | [1 0;1 1] | 94.4 | 90.24 |
| *m3* | [2 0;0 2] | [1 1;1 1] | [1 1;1 1] | 93.96 | 93.57 |
| *m4* | [2 0;0 2] | [2 2;3 3] | [1 1;1 1] | 87.67 | 86.89 |
| *m5* | [4 0;0 4] | [1 2;2 1] | [1 1;1 1] | 88.64 | 75.19 |
| *m6* | [4 1;1 4] | [2 1;1 2] | [1 0;0 1] | 92.18 | 83.45 |
| *m7* | [5 0;0 5] | [1 1;2 2] | [1 1;1 1] | 93.51 | 83.67 |
| *m8* | [5 1;1 5] | [1 2;2 1] | [1 1;1 1] | 91.56 | 64.8 |
| *m9* | [5 2;2 5] | [3 3;3 3] | [1 2;2 1] | 95.31 | 87.54 |

$$n_L(t) = f[n_L(t-1), W_f(t-1), A_8(t-2)] \tag{2.75}$$

where the output order $n_a$, the input order $n_b$, and the output delay $n_k$ are:

$$n_a = \begin{bmatrix} 1 & 0 \\ 0 & 1 \end{bmatrix}, n_b = \begin{bmatrix} 1 & 1 \\ 1 & 1 \end{bmatrix}, n_k = \begin{bmatrix} 1 & 2 \\ 1 & 2 \end{bmatrix}. \tag{2.76}$$

***Model Validation***

With the above parameters, change fuel flow and nozzle area to different input signals to further validate the identified *m1* for NARX-1. By comparing *m1* outputs with those from the component-level model, the error signals are given in Fig. 2.22. It is seen that the error for $n_H$ is well within 0.15% while for $n_L$ within 2.2%. The accuracy of *m1* is clearly validated.

***Development of NARX-2 and PNARX***

With the above procedures for data preparation, model parameters choice, and model identification in NARX-1, NARX-2 (90% $n_{H\max}$ ~ 100% $n_{H\max}$) can be established, see Table 2.7.

The model *n2* shows a very high accuracy in both $n_H$ and $n_L$ with best fitting 95.71 and 89.05%, respectively. It is thus chosen as the model for NARX-2 with representations as:

$$n_H(t) = g[n_H(t-1), W_f(t-1), A_8(t)] \tag{2.77}$$

$$n_L(t) = g[n_L(t-1), W_f(t-1), A_8(t-1), A_8(t-2), A_8(t-3)] \tag{2.78}$$

where the NARX-2 output order $n'_a$, the input order $n'_b$, and the output delay $n'_k$ are:

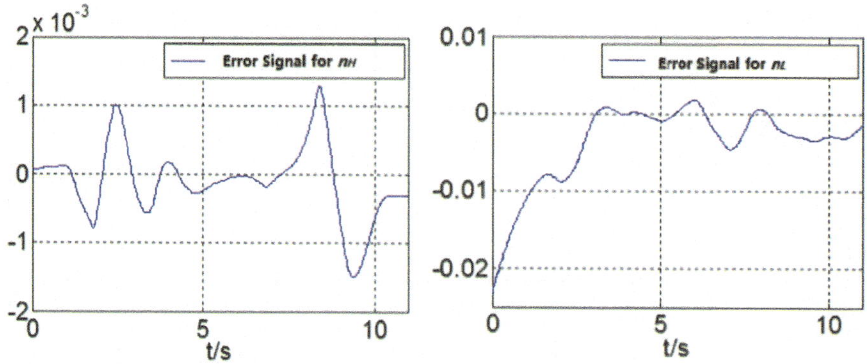


(a) (b)

**Fig. 2.22** Model validation for $m1$—for different input signals of $W_f$ and $A_8$ for validating the dynamic performance of NARX-1 model, **a** denotes the error signal between component-level output and $m1$ for $n_H$; and **b** is the error signal between component-level output and $m1$ output for $n_L$

**Table 2.7** Different parameters for NARX-2 Model (90% $n_{H\max}$ ~ 100% $n_{H\max}$)

| No | $n'_a$ | $n'_b$ | $n'_k$ | $n_H$(Fit %) | $n_L$(Fit %) |
|---|---|---|---|---|---|
| *n1* | [1 0;0 1] | [1 1;1 1] | [1 1;1 1] | 94.9 | 86.05 |
| *n2* | [1 0;0 1] | [1 1;1 3] | [1 0;1 1] | 95.71 | 89.05 |
| *n3* | [2 0;0 2] | [1 1;1 1] | [1 1;1 1] | 71.43 | -48.03 |
| *n4* | [2 0;0 2] | [2 2;3 3] | [1 1;1 1] | 86.52 | 77.28 |
| *n5* | [3 1;1 3] | [1 1;1 1] | [1 1;1 1] | 69.01 | −57.5 |
| *n6* | [4 0;0 4] | [1 2;2 1] | [1 1;1 1] | 73.3 | 79.48 |
| *n7* | [5 0;0 5] | [1 1;2 2] | [1 1;1 1] | 84.11 | 81.27 |
| *n8* | [5 1;1 5] | [1 2;2 1] | [1 1;1 1] | 80.44 | 67.32 |
| *n9* | [5 2;2 5] | [3 3;3 3] | [1 2;2 1] | 87.7 | 72.33 |

$$n'_a = \begin{bmatrix} 1 & 0 \\ 0 & 1 \end{bmatrix}, n'_b = \begin{bmatrix} 1 & 1 \\ 1 & 3 \end{bmatrix}, n'_k = \begin{bmatrix} 1 & 0 \\ 1 & 1 \end{bmatrix}. \tag{2.79}$$

Now changing input signals to further validate the identified *n2* for NARX-2, a comparison of *n2* outputs with those from the component-level model shows that the error in $n_H$ is within 0.5% while for $n_L$ within 1.5%. Thus, *n2* as the model for NARX-2 is validated.

With NARX-1 and NARX-2, the PNARX model is finally established with the rotational speed as the scheduling variable. The block diagram for validation of PNARX model is shown in Fig. 2.23. Although not shown here but from idle to TO, the PNARX model demonstrates an accuracy of 1.5% in $n_H$ and 4.5% in $n_L$.

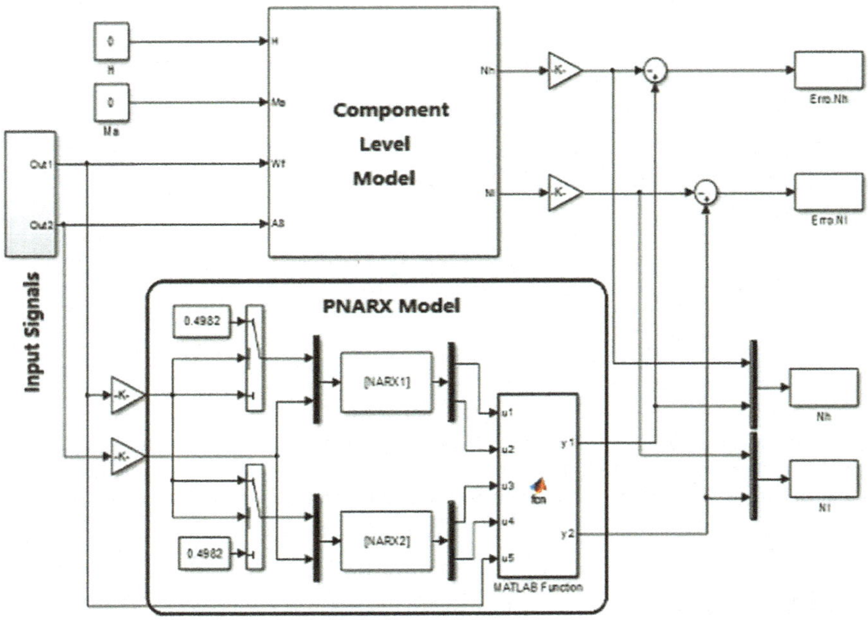


**Fig. 2.23** Block diagram for validation of PNARX model

Consequently, it can be utilized for initial investigation for either advanced control or EHM purposes.

## 2.8 Comparison for Aeroengine Nonlinear Identification Models

With the above development, a comparison is made in Table 2.8. The modeling accuracy is clearly shown with IBPNN demonstrating a very high degree of fit in both high-pressure rotational speed and low-pressure rotational speed. In specific, the following observations are given:

**Table 2.8** Model comparison

| | BPNN | IBPNN | PNARX |
|---|---|---|---|
| Inputs | $W_f$, $A_8$ | $W_f(t)$, $A_8(t)$ *and Regressors* | $W_f(t)$, $A_8(t)$ *and Regressors* |
| Outputs | $n_H, n_L$ | $n_H, n_L$ | $n_H, n_L$ |
| $n_H$ error (%) | 2.3 | 0.2 | 1.5 |
| $n_L$ error (%) | 11 | 2 | 4.5 |
| Model No (%) | 1 | 1 | 2 |

(1) BPNN is a static mapping between inputs and outputs, and thus suitable for steady state modeling used for, e.g. performance analysis etc.;
(2) For dynamic modeling, both IBPNN and PNARX models can be developed, which can capture both steady state and dynamic properties of turbofan engines;
(3) Although both IBPNN and PNARX models can be used for advanced control and engine health management purposes, IBPNN is essentially not a control-oriented modeling method, and PNARX model is recommended for this purpose;
(4) To achieve high modeling accuracy, piece-wise scheduling approach can be used. However, simulation shows that PNARX needs a greater number of sub-models to "piece" together than IBPNN. This can lead to implementation complexity in practice.

In summary, different methods exist for engine modeling, and a specific choice depends on the purpose of modeling, e.g. analysis of steady state property or dynamic property; for control design purpose, engine health management purpose or both. In cases where different modeling methods can satisfy the requirement, implementation issues have to be considered for further refinement of investigation. Although the identified models are mostly used in engine control units, an increasing number of sub-models often lead to complicated switching logic that should be better avoided. Thus, a balanced consideration should be hold for engine modeling.

## 2.9 Summary

Aeroengine component-level modelling and identification-based modelling have been presented. The former can take heat transfer, time delay, turbine cooling ect, thus can be very accurate (temperature, pressure and velocity within 1%); but component-level models usually have poor real time performance; with EEC and sensor technologies, real time component-level models can be developed for on-board control and health management, but they are essentially not control-oriented. This necessitates "control-oriented on-board real-time models", which are usually obtained through identification methods.

The principle of identification-based models is to treat the engine as a "black box", and then "guess" the engine dynamics through experimental data! System Identification has been a wide yet independent subject, and systematic methods have been developed, e.g. Least Squares, Generalized Least Squares, Instrumental Variables, Prediction Error Methods, Maximum Likelihood, as well as many recursive algorithms etc. Identification-based methods can obtain both "control-oriented on-board real-time models" and "fault detection-oriented models", which is at the core of advanced control and health management.

These form the basis for carrying out advanced control designs as most of the control design methods assume a model of the engine existing. However, it is noted that different models are also required as control systems possess different functions.

For each function, control design requires different model representations. These will be seen clearly as in the following chapters.

## References


1. Walsh, P. P., and P. Fletcher. 2004. *Gas turbine performance,* 2nd ed. USA: Blackwell Science Ltd. Print ISBN 9780632064342|Online ISBN 9780470774533|DOI https://doi.org/10.1002/9780470774533
2. Yao, H. *Full authority digital electronic control system for aero-engine.* Beijing: AVIC Publishing Ltd.
3. Li, Y.G., P. Pilidis, and M.A. Newby. 2006. An adaptation approach for gas turbine design point performance simulation. *ASME Journal of Engineering for Gas Turbines and Power* 128: 789–795.
4. Lichtsinder, M., and Y. Levy. 2002. Two-sprool, turbo-fan engine: (a) steady-state and dynamic mathematical models. In *TAE No. 903*, Technion, Israel, 20–37.
5. Khalid, S. J., and R.N. Hearne. 1980. Enhancing dynamic model fidelity for improved prediction of turbofan engine transient performance. *AIAA* 80–1083.
6. Lichtsinder, M., and Y. Levy. 2006. Jet engine model for control and real-time simulations. *ASME Journal of Engineering for Gas Turbines and Power* 128: 745–753.
7. Martin, S., I. Wallace, and D. G. Bates. 2008. Development and validation of a civil aircraft engine simulation model for advanced controller design. *ASME Journal of Engineering for Gas Turbines and Power* 130 (5): 051601:1–15.
8. Reed, J. A., and A. A. Afieh. 1999. Computational simulation of gas turbines PART I-foundations of component-based models. In *Proceedings of the International Gas Turbine & Aeroengine Congress & Exhibition, Indianapolis*, Indiana, 7–10 June 1999.
9. Sanghi, V., B.K. Lakshmanan, and V. Sundararajan. 2012. Survey of advancements in jet-engine thermodynamic simulation. *Journal of Propulsion & Power* 16 (5): 797–807.
10. Martin, S., I. Wallace, and D.G. Bates. 2008. Development and validation of a civil aircraft engine simulation model for advanced controller design. *Proceedings of American Control Conference* 130 (5): 2334–2339.
11. Liu, N., J. Huang, F. Lu, and M. Pan. 2015. Frequency domain identification of multivariable model for aeroengine using an improved maximum likelihood method. *International Journal of Turbo & Jet Engines* 32.
12. Pakmehr, M. 2013. Towards verifiable adaptive control of gas turbine engines. PhD Dissertation, Georgia Institute of Technology.
13. Behbahani, A., E. Feron, J. Paduano, N. Fitzgerald, and M. Pakmehr. 2011. Decentralized piecewise linear modeling of a turboshaft engine driving a variable pitch propeller. *AIAA Journal.*
14. Csank, J. 2010. Control design for a generic commercial aircraft engine. *AIAA Journal.*
15. Austin Spang III, H., and H. Brown. 1999. Control of jet engines. *Control Engineering Practice,* 7: 1043–1059.
16. Asgari, H., M. Venturini, X. Q. Chen, and R. Sainudiin. 2014. Modeling and simulation of the transient behavior of an industrial power plant gas turbine. *ASME Journal of Engineering for Gas Turbines & Power* 136 (6).
17. Wang, J., Z. Ye, and Z. Hu. 2014. Onboard real time modeling of aircraft engines with a Hammerstein-Wiener representation. *Journal of Aerospace Power* 29 (10): 2499–2506.
18. Lichtsinder, M., and Y. Levy. 2006. Jet engine model for control and real-time simulations. *ASME Journal of Engineering for Gas Turbines & Power* 128: 745–753.
19. Embrechts, M. J., A. L. Schweizerhof, M. Bushman, and M. H. Sabatella. 2000. Neural network modeling of turbofan parameters. ASME Paper No. 2000-GT-0036.

20. Zhuo, G., J. Sun, and G, Yang. 2004. Aeroengine modeling based on wavelet neural network. *Journal of Nanjing University of Aeronautics & Astronautics* 36 (6): 728–731.
21. Liu, Y., H. Xing, and S. Huang. 2012. Adaptive simulation of gas turbine performance using improved genetic algorithm. *Journal of Aerospace Power* 27 (3): 695–700.
22. Irwin, G. W., K. Warwich, and K. J. Hunt (eds.).1995. Neural network applications in control. *IET Digital Library*. e-ISBN 9781849193498.
23. Ljung, L. 1999. *System identification—Theory for the user*, 2nd ed. Upper Saddle River, N.J.: PTR Prentice Hall.
24. Schoukens, J., J. Suykens, and L. Ljung. 2009. Wiener-Hammerstein benchmark (SYSID 2009 special session).
25. Eskinat, E., S. H. Johnson. 1991. Use of Hammerstein models in identification of nonlinear systems. *AI.Ch.E. Journal* 37: 255–268.
26. Bai, E., and M. Fu. 2002. A blind approach to Hammerstein model identification. *IEEE Transactions on Acoustics, Speech and Signal Processing* 50: 1610–1619.
27. Vörös, J. 2004. An iterative method for Hammerstein-Wiener systems parameter identification. *Journal of Electrical Engineering* 55: 328–331.
28. Crama, Ph., and J. Schoukens. Hammerstein-Wiener system estimator initialization. *Proceedings of ISMA2002* III: 1169–1176.
29. The User's Guide to Matlab, Mathworks Incorp.
30. Samar, R. 1995. Robust multi-mode control of high performance aeroengines. PhD Thesis, the University of Leicester.
31. Sutton, A. E. 1992. The application of multivariable control to a turbofan engine. Technical Report TMP1220, Defence Research Agency, Aerospace Division, Farnborough, Hampshire, UK.
32. Greig, A. W. M. 1994. Multivariable powerplant control-proof of concept. Technical Report DRA/AS/PTD/TR/94067/1, Defence Research Agency, Farnborough, Hampshire, UK.
33. Eisa, S. A., and H.P. Tyler. 1986. Closed loop control of an afterburning F100 gas turbine engine. In *Proceedings of the American Control Conference*, Seattle, WA, USA, 266–272.
34. Jaw, L. C., and J.D. Mattingly. 2009. *Aircraft engine controls: Design, system analysis and health monitoring*. Virginia: AIAA, Inc.
35. Xia, C., J. Wang, G. Shang, M. Zhou. 2012. Component-level modeling and analysis of aeroengine based on Matlab/Simulink. *Aeroengine* 38 (4): 31–33, 52 (In Chinese).
36. Wang, J., Z. Ye, and Z. Hu. 2012. Nonlinear control of aircraft engines using a generalized Gronwall-Bellman lemma approach. *ASME Journal of Engineering for Gas Turbines and Power* 134(9): 094502.

# Chapter 3
# Model-Based Aeroengine Nonlinear Set Point Control

Aeroengine control systems consist of fuel pump systems, burner and afterburner fuel metering devices, bleeding valves, variable geometry actuators, sensors, tip clearance controls, generators and engine control units etc., to fulfil but not limited with functions such as power management, start logic, idle control, takeoff control, climbing control, fuel metering, acceleration and deceleration control, limit protection control, variable stator vane control, valve bleeding control, tip clearance control, anti-surge control, together with monitoring, fault detection and other intelligent functions etc. From aeroengine control systems requirements perspective, it mainly includes functional requirement as state control and state monitoring; performance requirement as steady state performance, dynamic performance, and switching performance; environment requirement as components operational conditions; as well as "warrants requirements" as safety, reliability, maintainability, supportability, provability.

From a design perspective, structure and function must be fulfilled simultaneously—structures determine functions. Thus, it is seen that aeroengine control systems can be complicated due to the various functions they must provide; meanwhile from system engineering perspective, they include control system overall design, components design and selection, control software design etc. Control system overall design are determined through comparative analysis among the proposals, in terms of system requirements. It mainly includes main loop design as fuel loop control, guide vane control, nozzle control and vector nozzle actuator control (for military engines) etc.; control system redundancy design as double redundancy, double channels, and reliability analysis; control system accuracy design as control accuracy vs. control plan, sensors vs. actuators; stability and stability margin design; primary component principle and parameters design. Based on the above designs, further refinements for EEC, fuel systems, control software, sensors function/performance parameters are to be developed.

This roughly summarizes the research and development process for aeroengine control systems. Yet, to simplify the discussion, it is focused on aeroengine primary loop control, whose design procedures usually follow the following steps:

 
J. Wang et al., *Model-based Nonlinear Control of Aeroengines*,
https://doi.org/10.1007/978-981-16-4453-5_3

(1) Divide flight envelope into regions;
(2) Approximate by a linear model within each region;
(3) Design controllers using linear design—**sep point controllers**;
(4) Design accel-decel schedules—**transient controllers**;
(5) Use gain scheduling to fulfil safe and efficient control over full flight envelope—**limit protection controllers**.

That is, control designs are functionally divided into the following three basic controls:

(1) **Aeroengine set point control**: Determine desired operation condition based on the pilot's command—throttle position or a PLA, and then regulate engine performance in the neighbourhood of this desired operation condition.
(2) **Aeroengine transient control**: Fulfil acceleration and deceleration while providing good transient performance.
(3) **Aeroengine limit protection control**: For both set point control and transient control, aeroengine should be protected from running beyond operating limits.

Aeroengine as a safety–critical system, it must be protected from running beyond any limits including:

(1) engine protection from undesirable operating conditions like stall, surge, Exceeding the maximum LP-shaft physical speed limit (afterburner nozzle area opens too fast or flame out);
(2) Exceeding the maximum HP-shaft physical speed limit (LP-shaft goes into overspeed for whatever reason);
(3) Exceeding the maximum compressor exist static pressure limit (engine overspeed or inlet pressure exceeds design limit);
(4) Exceeding the maximum EGT limit in turbines (engine fuel-rich or excessive extraction of bleed air or mechanical power);
(5) Exceeding the maximum compressor exit temperature limit (engine overspeed or inlet temperature exceeds design limit);
(6) Other limits, e.g. afterburner light-off limit, engine air start limit, structural limits etc.

The above three basic functions can be clearly explained by referring to Fig. 3.1 on a compressor map, where the three shaded circles represent the operating area for set point controllers; the dotted lines show the acceleration and deceleration procedures; and finally, the different limits indicate that they must be respected for any condition over the flight envelope. The remaining chapters will centre on the three basic functions, first introducing their "standard practice," and then moving into advanced design methods particularly model-based nonlinear controls.

The current chapter will first focus upon aeroengine set point control. It is seen that the function of set point control is to regulate the engine's performance near a desired operating condition such as idle, cruise or takeoff. Therefore, from a control design perspective, set point control is essentially a "**disturbance rejection problem**" and the resulting design should provide fast regulation of the engine, e.g. be able to reject disturbances and robust against modelling uncertainties.

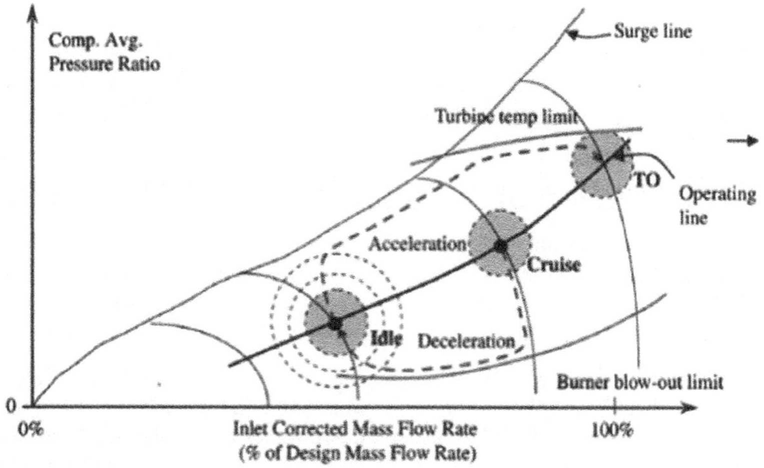


**Fig. 3.1** Three basic functions as referred on a compressor map

The disturbance rejection problem is well known in both linear control and nonlinear control theories, e.g. the celebrated unified framework of $H_2$ and $H_\infty$ control for linear systems [1–4]; self-tuning regulator adaptive control for nonlinear systems [5, 6] etc. In aeroengine control industry, as mentioned before, PID or LQR are usually adopted where the design proceeds with a linear model. Henceforth conventional set point controllers are also called "model-based linear design."

Nonlinear systems theory has developed rapidly over recent decades including concepts such as zero dynamics and normal forms [7], passivity and dissipativity [8], nonequilibrium theory [9] etc. As a consequence, a number of nonlinear control design techniques have been well established such as feedback linearization [7], recursive designs including backstepping and forwarding [10], energy-based control design for nonholonomic dynamical systems [11] and nonlinear model predictive control [12], to name just a few. One of the central themes in nonlinear control is to achieve robustness through feedback design, that is, acceptable performance should be retained even under possibly severe disturbances.

Indeed, several approaches have been developed to handling uncertainties, the well-known H$\infty$ control approach (linear [13], nonlinear [14], piecewise linear [15, 16]) deals with uncertainty as bounded disturbance and the optimal controllers are such designed to meet performance requirement even for the worst-case scenario. This will inevitably introduce conservatism into the design and one way to avoid this is through disturbance observation—disturbance observer. This approach is powerful since the disturbance can be estimated so that "prognostic" action can be taken to reduce the conservatism. Therefore, a further design freedom can be added in order to counteract the usually detrimental effect of disturbances in conjunction with feedback control [17]. It is worth mentioning that there are various forms of disturbance observers [18–20] which have found many practical applications (e.g. [21–23]). One of the disturbance observer approaches to robust control is active disturbance rejection control (ADRC) proposed by Han [24], consisting of tracking differentiation, extended state observer and a nonlinear PD. ADRC thus achieves robust control by compensating nonlinearity, model uncertainly and exogenous disturbances in real time.

Yet, a literature survey shows that the rich development in nonlinear control theories has not been taken advantage of in aeroengine control designs. The investigation of model-based nonlinear control is necessitated, apart from practical motivations mentioned before. In particular, Sect. 3.1 introduces a generalized Gronwall-Bellman lemma-based approach where the convergence bound can be estimated; a control Lyapunov function-based design methodology is then introduced to provide an a priory stability guarantee in Sect. 3.2; joint design with generalized Gronwall-Bellman lemma and Lyapunov function is subsequently proposed to provide fast transition of aeroengine with stability in Sect. 3.3. It is seen that usually the Lyapunov function methods require full-state feedback and it is natural to ask if it is possible to have partial state feedback control, e.g. in the simplest case, single sensor, single actuator static state feedback, and this problem is addressed in Sect. 3.4; to further enhance the regulation time, a finite-time version of the restricted Lyapunov function method is proposed for significantly improve the set point control performance in Sect. 3.5 before Sect. 3.6 summarizes this chapter on set point control.

## 3.1 Generalized Gronwall-Bellman Lemma Approach for Set Point Control

As having been indicated, set point can has been conventionally designed using linear control design methodologies such as PID control, LQR/LTR [25] or H$\infty$ optimal control [26–28], with almost exclusively of PID control in practice. But nonlinear control techniques are indeed required and this section introduces a new nonlinear design method based on a generalized Gronwall-Bellman lemma approach. This work has been motivated by the authors' effort to extend the control envelope around the operating point, in the hope that through such an extension, performance can be improved together with plant variation accommodation. Putting in the context of aircraft engine control, one trend is to refine the linear model around a particular operating point while the other research direction is to base control design on model identification feasible over the full flight envelope. Linear model refinement at a certain flight condition is desirable since this may result in improved performance, but robustness becomes a demanding issue due to large environmental changes that soon challenge the accuracy of the refined linear model; on the other hand, from the author's point of view, it seems too ambitious that one single controller can achieve acceptable performance over the full flight envelope. As a consequence, considering a nonlinear model can represent the engine dynamics over a fairly large envelope, it is not unreasonable to pursue the corresponding nonlinear design techniques. This inevitably necessitates the employment of the mechanism from nonlinear systems theory. The result presented here is thus an attempt towards the above idea. The roadmap is as follows: the theoretical preliminaries are presented in Sect. 3.1.1; the main result on the nonlinear control of aircraft engines is studied in Sect. 3.1.2; then Sect. 3.1.3 provides a case study for the validation of the proposed control design method; and finally Sect. 3.1.4 summarizes this section.

### *3.1.1 Preliminaries*

The Gronwall-Bellman lemma has long been used for the stabilization of nonlinear systems (see [29–31] for example) as well as nonlinear observer design [32]. In its "standard" form, the Gronwall-Bellman lemma can be stated as follows:

**Gronwall-Bellman Lemma [29–31]** *Let the following conditions hold:*

(i) *$f$, $g$and $k(\Re^+ \mapsto \Re)$ are locally integrable, where $g \geq 0, k \geq 0$and $g \in \ell\infty$;*
(ii) *$gk$is locally integrable on $\Re^+$;*

*If $u : \Re^+ \mapsto \Re$satisfies*

$$u(t) \leq f(t) + g(t) \int_0^t k(\tau)u(\tau)d\tau, \quad \forall t \geq 0 \tag{3.1}$$

then

$$u(t) \leq f(t) + g(t) \int_0^t k(\tau) f(\tau) \exp\left( \int_\tau^t k(s)g(s)ds \right) d\tau, \quad \forall t \geq 0 \tag{3.2}$$

The above lemma is very useful for the analysis of exponential stabilization. For example, take $f \equiv c$ and $g \equiv 1$ leads to:

$$u(t) \leq c + c \int_0^t k(\tau) \exp\left( \int_\tau^t k(s)ds \right) d\tau, \quad \forall t \geq 0 \tag{3.3}$$

After an equivalent manipulation resulting in:

$$u(t) \leq c \exp\left( \int_0^t k(\tau)d\tau \right), \quad \forall t \geq 0 \tag{3.4}$$

which admits exponential behaviour of $u(t)$.

However, for the purpose of this paper, the basic Gronwall-Bellman lemma must be further generalized. One of the most important generalizations, pertaining to the main results of the current paper, has been proposed by Pachpatte [33, 34], El Alami [35] and more recently, N'Doye et al. [36].

**Generalized Gronwall-Bellman Lemma [35, 36]** *let*

(i) $a,b,k \in \Re$ *with* $0 \leq a < b$ *and* $k > 0$; *further define an integer* $l > 1$;
(ii) $f:\Re^+ \mapsto \Re^+$ *an integrable function with* $\int_\alpha^\beta f(s)ds > 0, \forall \alpha, \beta \in [a, b], 0 \leq \alpha < \beta$;
(iii) $x : [a, b] \mapsto \Re^+$ *an essentially bounded function such that:*

$$x(t) \leq k + \int_a^t f(s)[x(s)]^l ds, \quad \forall t \in [a, b] \tag{3.5}$$

Then if the following inequality

$$1 - (l-1)k^{l-1} \int_a^b f(s)ds > 0 \tag{3.6}$$

holds, then one has:

$$x(t) \leq \frac{k}{\left(1 - (l-1)k^{l-1} \int_a^t f(s)ds\right)^{\frac{1}{l-1}}}, \forall t \in [a, b] \tag{3.7}$$

It is seen that bounding of $x(t)$ can be achieved under appropriate conditions. Such a generalization to the Gronwall-Bellman lemma will be taken as the starting point for control design. This is delineated in the next section.

### *3.1.2 Nonlinear Set Point Control of Aeroengines: Theoretical Results*

A nonlinear model of an aircraft engine can be represented as:

$$\begin{aligned} \dot{x} &= f(x, u) \\ y &= g(x, u) \end{aligned} \tag{3.8}$$

A state space model can be obtained at a certain steady state through Taylor's series expansion, resulting in:

$$\begin{aligned}\Delta\dot{x} &= A\Delta x + B\Delta u \\ \Delta y &= C\Delta x + D\Delta u\end{aligned} \tag{3.9}$$

Consider, for example, fuel flow ratio $W_f$ as the control input, rotation speeds of the compressor and fan, $n_H$ and $n_L$, as the state variables, then the state equation can be written as:

$$\begin{bmatrix}\Delta\dot{n}_L \\ \Delta\dot{n}_H\end{bmatrix} = \begin{bmatrix}a_{11} & a_{12} \\ a_{21} & a_{22}\end{bmatrix}\begin{bmatrix}\Delta n_L \\ \Delta n_H\end{bmatrix} + \begin{bmatrix}b_1 \\ b_2\end{bmatrix}\Delta W_f \tag{3.10}$$

The parameters of the equation are then obtained through the small perturbation method.

It is obvious that such a model can only be valid around the steady state operating point. Therefore, it is desirable to extend the feasibility of the model over a larger envelope. One of the strategies can be sought to augment the model with nonlinear terms, e.g.:

$$\Delta\dot{x} = A\Delta x + G(\Delta x) + B\Delta u \tag{3.11}$$

where $G(\bullet)$ can be a very general nonlinear term.

In fact, consider the generality of the expression for $G$ as well as for ease of exposition, Eq. (3.11) is rewritten as:

$$\dot{x} = Ax + G(x) + Bu \tag{3.12}$$

It is thus hoped that through such an augmentation of the system plant, although now it becomes nonlinear necessitating the nonlinear design technique, the control performance can be improved even over a large flight envelope. In the following, nonlinear control of aircraft engines will be analysed through the generalized Gronwall-Bellman lemma approach.

To proceed, consider the static state feedback control $u = Kx$, and the problem of nonlinear control design is to find a constant gain $K$, such that the state $x(t)$ is stabilized subject to the constraint:

$$\dot{x} = Ax + G(x) + BKx \tag{3.13}$$

together with initial condition $x_0$.

Before we present the main result, the following assumptions are made:

**Assumption** The nonlinear dynamical system (12) satisfies the following conditions:

(i) the pair $\left(A,\ B\right)$ is stabilizable;
(ii) there exists an integer $q \geq 1$, such that

$$\|G(x(t))\| \leq \gamma \|x(t)\|^q \tag{3.14}$$

where $\gamma$ is a positive constant that can be a design parameter tuning the exponential behaviour.

In the above assumption, (i) is standard while (ii) essentially says that $G(x)$ is bounded and Lebesgue measurable, which is true for aircraft engines. The main result can now be presented as below.

**Theorem (Nonlinear Control of Aircraft Engines)** *Under the above assumption, the system (12) controlled by the state feedback control*

$$u = Kx \tag{3.15}$$

*is exponentially stable if all the eigenvalues of matrix* $(A + BK)$*have a strictly negative real part with the initial condition satisfies:*

$$\|x_0\|^{q-1} < \frac{|\lambda|}{\gamma M^q} \tag{3.16}$$

*where the constants* $M > 0$*and* $\lambda < 0$*is specified by:*

$$\left\|e^{(A+BK)t}\right\| < Me^{\lambda t}, \quad \forall t \geq 0. \tag{3.17}$$

*In fact the state* $x(t)$*is explicitly bounded by:*

$$\|x(t)\| \leq \frac{M\|x_0\|e^{\lambda t}}{\left(1 - \frac{\gamma M^q \|x_0\|^{q-1}}{|\lambda|}\right)^{\frac{1}{q-1}}}. \tag{3.18}$$

***Proof*** Consider the dynamical system (13), with assumption 1 the differential equation can be integrated to be:

$$x(t) = e^{(A+BK)t}x_0 + \int_0^t e^{(A+BK)(t-s)}G(x)ds \tag{3.19}$$

Hence:

$$\|x(t)\| \leq Me^{\lambda t}\|x_0\| + Me^{\lambda t}\int_0^t \gamma e^{-\lambda s}\|x\|^q ds \tag{3.20}$$

Or:

$$\|x(t)\|e^{-\lambda t} \leq M\|x_0\| + \gamma M\int_0^t e^{(q-1)\lambda s}\left(\|x\|e^{-\lambda s}\right)^q ds \tag{3.21}$$

Now it is time to invoke the generalized Gronwall-Bellman lemma, where it can be easily checked that the conditions for the lemma to hold are fulfilled. Hence, provided that

$$1 - (q-1)(M\|x_0\|)^{q-1}\int_0^t \gamma Me^{(q-1)\lambda s}ds > 0 \tag{3.22}$$

the behaviour of $\|x(t)\|$ is bounded by:

$$\|x(t)\|e^{-\lambda t} \leq \frac{M\|x_0\|}{\left(1 - (q-1)(M\|x_0\|)^{q-1}\int_0^t \gamma Me^{(q-1)\lambda s}ds\right)^{\frac{1}{q-1}}} \tag{3.23}$$

The inequality (21) can be further reduced to:

$$1 - \frac{\gamma M^q\|x_0\|^{q-1}}{|\lambda|}\left[1 - e^{(q-1)\lambda t}\right] > 0 \tag{3.24}$$

while (22) leads to:

$$\|x(t)\| \leq \frac{M\|x_0\|e^{\lambda t}}{\left(1 - \frac{\gamma M^q\|x_0\|^{q-1}}{|\lambda|}\left[1 - e^{(q-1)\lambda t}\right]\right)^{\frac{1}{q-1}}} \tag{3.25}$$

A consideration of the inequality $1 - e^{(q-1)\lambda t} < 1$ gives the desired results. □

**Remark 1** It is remarkable that the behaviour of $\|x(t)\|$ can be "tuned" by the scalar constants $M$,$\lambda$, $\gamma$ and $q$, which are subsequently derived from $\left\|e^{(A+BK)t}\right\| < Me^{\lambda t}$ and $\|G(x(t))\| \leq \gamma\|x(t)\|^q$.

**Remark 2** For the design purpose, $\gamma$ and $q$ can be easily estimated from the nonlinear term $G(x)$; then a choice of static gain $K$ will determine $M$ and $\lambda$, announcing the fate of $\|x(t)\|$ in terms of inequality (18), gives a flavour of a priori design.

### 3.1.3 Nonlinear Set Point Control of Aeroengines: Numerical Study

#### *Regulation Design*

The linear part of the engine model used here is adopted from reference [27] that has been constructed in the Intelligent Engine Control project [37]. The nonlinear function $G(x)$ is assumed to be:

$$G(x) = \begin{bmatrix} 12x_1^2 - x_2^2 \\ -1.7x_1^2 + x_2^2 \end{bmatrix} \tag{3.26}$$

representing the nonlinear correction to the nominal linear model. It is noted that there exists cross-channel interaction in $G(x)$. And such interaction can not be eliminated through coordinate transformation, indicating the complex dynamics of the aircraft engines.

The model is expressed as below:

$$\begin{aligned} \dot{x} &= \begin{bmatrix} -4.1476 & 1.4108 \\ 0.2975 & -3.1244 \end{bmatrix} x + G(x) + \begin{bmatrix} 0.2491 \\ 0.2336 \end{bmatrix} u \\ y &= \begin{bmatrix} 8.7379 & 0 \\ 3.3033 & 3.8052 \\ 2.1940 & 2.5749 \end{bmatrix} x \end{aligned} \tag{3.27}$$

where $u = \Delta W_f$ and $y^T = \begin{bmatrix} PCN2R & P56/P25 & P16/P56 \end{bmatrix}^T$.

When the engine works at steady state one has $u = \Delta W_f = 0$. Now consider a regulation problem, that is, the design objective is to design $u = \Delta W_f = Kx$ such that the transient behavior of the states $x(t)$ can be improved.

To proceed, as $\|G(x(t))\| \leq \gamma \|x(t)\|^q$, then one can choose $\gamma = 12$ and $q = 2$; the eigenvalues of the state equation are computed to be $\lambda_1 = -4.4615$ and $\lambda_2 = -2.8105$, now choose $K = \begin{bmatrix} k_1 & k_2 \end{bmatrix} = \begin{bmatrix} -5 & -7 \end{bmatrix}$ such that the eigenvalues of the controlled state equation are prescribed at $\lambda_1' = -5.70$ and $\lambda_2' = -4.45$, then from $\left\| e^{(A+BK)t} \right\| < Me^{\lambda t}$ one can designate $M = 1$ and $\lambda = -4.45$. Finally, the range of the initial conditions can be determined from inequality (16), $\|x_0\|^{q-1} < \frac{|\lambda|}{\gamma M^q}$, that is: $\|x_0\| < 0.371$. A simulation with initial condition $x_0 = \begin{bmatrix} -0.2 & -0.3 \end{bmatrix}^T$ (thus $\|x_0\| = 0.3606 < 0.371$), then the transient behaviour of $x(t)$ is shown in Fig. 3.2, also shown is the transient response of the uncontrolled state equation.

The evolution of $\|x(t)\|$ is shown in Fig. 3.3. It is seen clearly that the performance is significantly improved, cross-validating the performance improvement of the states $x_1(t)$ and $x_2(t)$ in Fig. 3.2. Meanwhile, the theoretical bound of the state $x(t)$ can be computed from inequality (18),

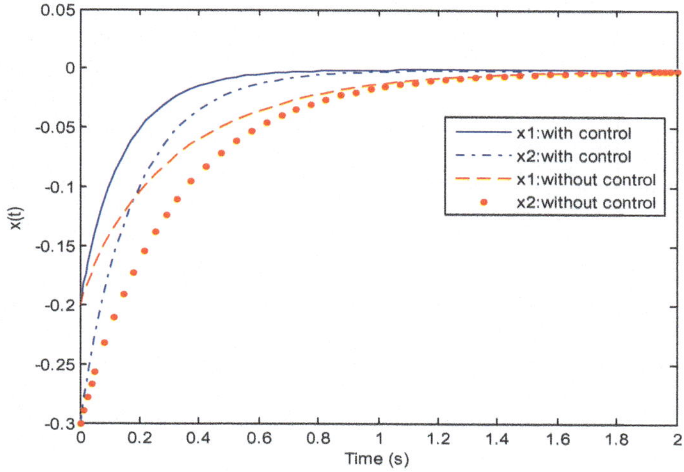


**Fig. 3.2** Comparison of transient response of state signals: $x_0 = [-0.2 \quad -0.3]^T$

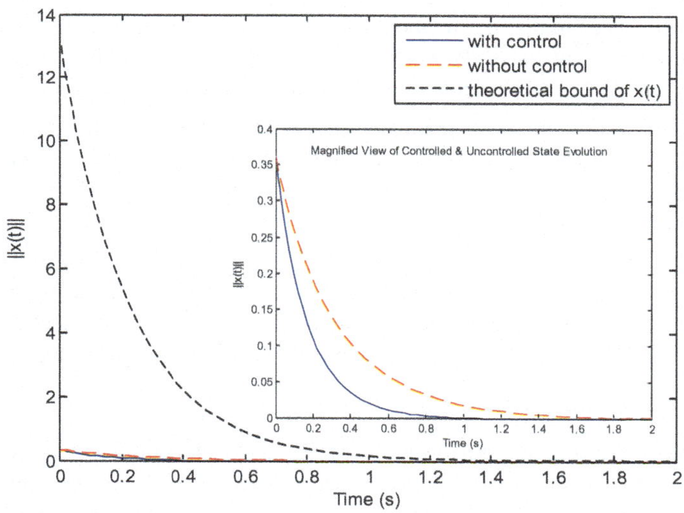


**Fig. 3.3** Comparison of transient response for $\|x(t)\|$ with theoretical bound: right corner is the magnified view of the state norm evolution for controlled and uncontrolled systems

$$\|x(t)\| \leq \frac{M\|x_0\|e^{\lambda t}}{\left(1 - \frac{\gamma M^q \|x_0\|^{q-1}}{|\lambda|}\right)^{\frac{1}{q-1}}} = \frac{\|x_0\|e^{-4t}}{(1 - 2.7\|x_0\|)}.$$

From the theorem, this says when the initial condition $\|x_0\| < 0.371$, then the state will be bounded from above by $\|x(t)\| \leq \frac{\|x_0\|e^{-4t}}{(1-2.7\|x_0\|)}$. This situation is also shown in Fig. 3.3, validating the theoretical result on the bounding behaviour of $\|x(t)\|$.

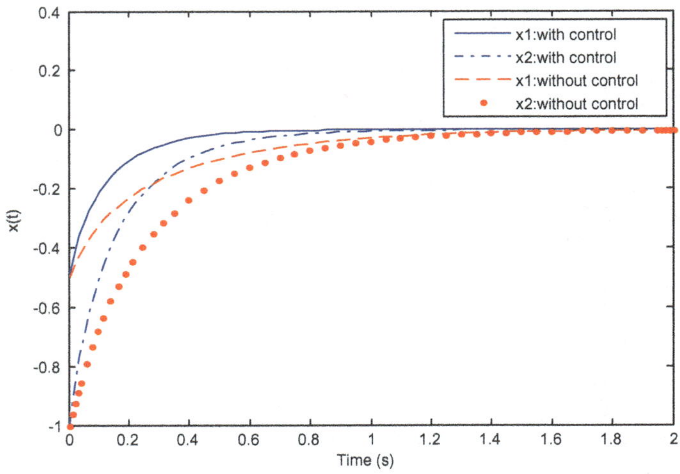


**Fig. 3.4** Comparison of transient response of state signals: $x_0 = [-0.5 \quad -1]^T$

Here comes an interesting question: observing Fig. 3.3, it might be argued that the evolution of state norm $\|x(t)\|$ is over-bounded and thus the range of initial conditions for $\|x_0\|$ is too restricted. And it is in fact interesting to show if the range can be extended. The answer is positive. Take, for example, $x_0 = \begin{bmatrix} -0.5 & -1 \end{bmatrix}^T$, thus $\|x_0\| = 1.118 > 0.371$, then with the same controller, the transient behaviour of $x(t)$ is shown in Fig. 3.4, also shown is the transient response of the uncontrolled state equation (open-loop response).

The regulation performance for $y^T = \begin{bmatrix} PCN2R & P56/P25 & P16/P56 \end{bmatrix}^T$ is shown in Fig. 3.5. It is demonstrated that the state feedback control has resulted in much improved performance. As a consequence, the engine can reject disturbance effectively, and this will be an appealing feature for aircraft engine control systems.

***Tracking Performance and Robustness***

Now consider the tracking performance of the above design: an increment of fuel flow ratio $\Delta W_f$ indicates acceleration process of the aircraft engine. This requires fast response of $PCN2R$ to the step input of $\Delta W_f$. For certain flight height $H$ and velocity *Ma*, this implies fast response of rotation speed (of high-pressure turbine for turbofan engines), which is extremely desirable as it is a measure of manoeuvrability of the aircraft. This situation is shown in Fig. 3.6. For comparison, the responses of the corresponding linearized systems are also shown in the figures. A detailed discussion follows:

(1) first consider the open loop and closed loop control of nonlinear systems: it is seen clearly that the resulting design significantly improves the transient response, yielding good tracking performance;

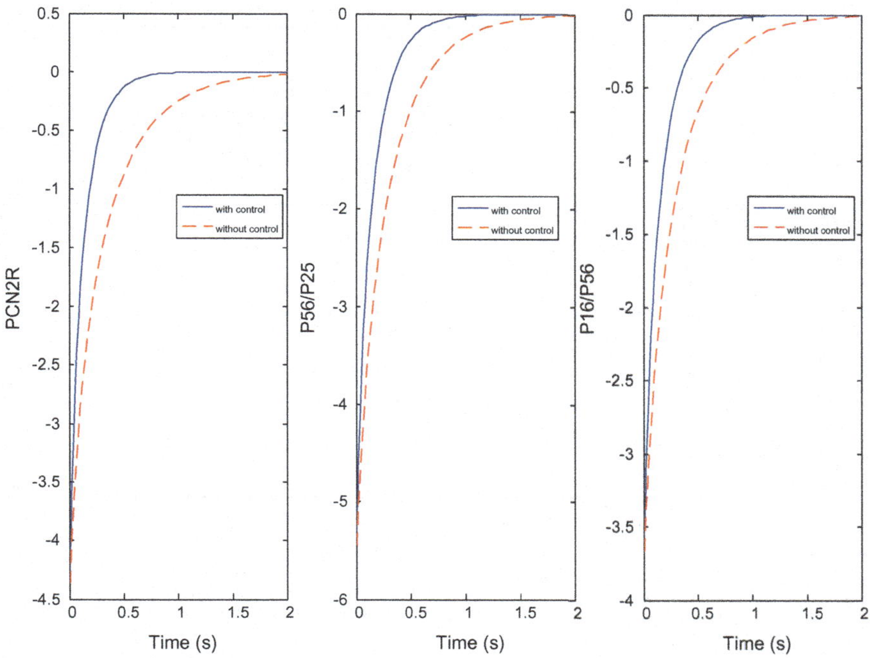


**Fig. 3.5** Comparison of regulation performance of output signals

(2) then consider the open loop and closed loop control of the linearized systems: the same state feedback controller also results in improved transient response, which can imply good tracking performance for a nominal design;

(3) now consider the open loop nonlinear system and the linearized system: it is observed that the proposed control design has achieved good tracking performance for both the original nonlinear system and its linearized counterpart, which indicates good robustness of the resulting design;

(4) finally consider the following scenario: the linearized system is the nominal plant for certain flight condition, and with the performance degradation, e.g. sealing leakage, blade erosion etc., the real system dynamics is now expressed by the nonlinear system, what can be drawn from Fig. 3.6 is that although the open loop performance has significantly deteriorated (this is indeed the case for real aircraft engine), the closed loop performance still maintains almost at the same level—in fact, from the comparison of the performance for closed loop nonlinear and linearized systems, it is seen that the lost of performance is marginal. This says that the proposed design can accommodate large plant variation and this is a desirable feature for control system.

To recap, the above simulation shows that the resulting design has achieved good performance for both disturbance regulation and fuel flow ratio tracking, validating

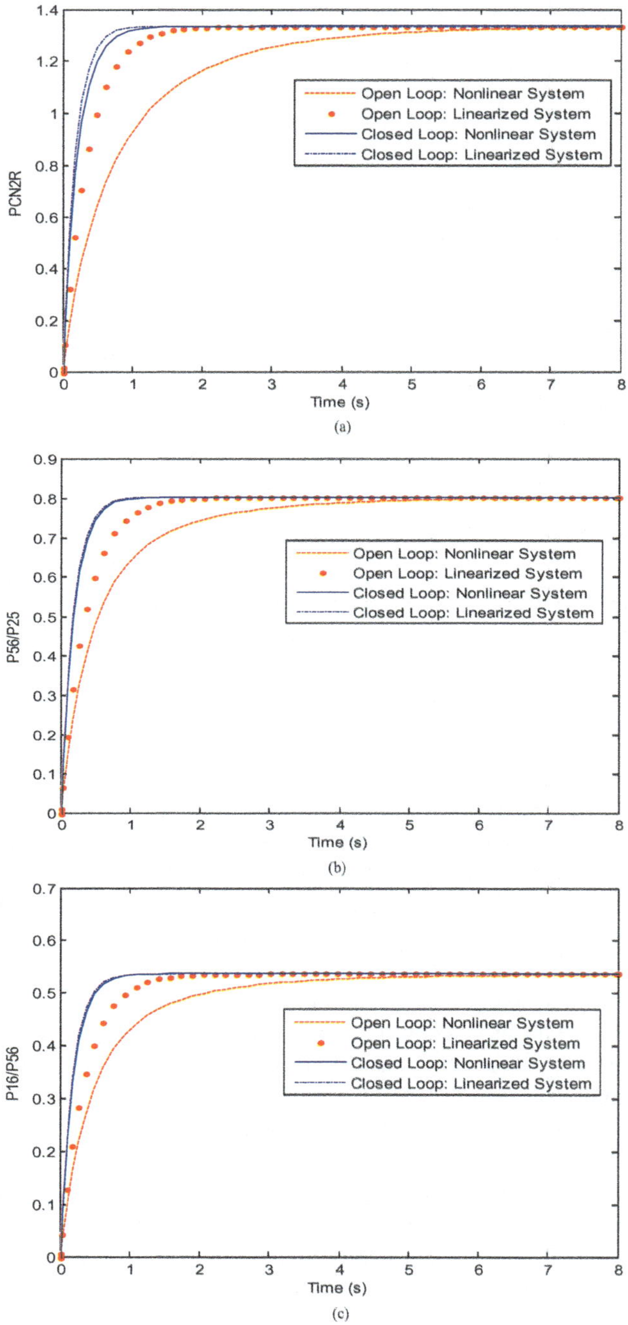


**Fig. 3.6** Tracking performance of the proposed design, also shown is the performance for corresponding linearized systems. **a** PCN2R. **b** P56/P25. **c** P16/P56

the effectiveness of the proposed design method based on the generalized Gronwall-Bellman lemma.

### 3.1.4 *Summary for Generalized Gronwall-Bellman Lemma Based Design*

This section has proposed a novel nonlinear design technique based on a generalized Gronwall-Bellman approach. Important theoretical results have been obtained on nonlinear control of aircraft engines. The theoretical results have also been validated by a numerical study, which has shown good performance for both regulation and tracking of the resulting design. The proposed method is easy to design and tuning with a promising potential application to nonlinear control of aircraft engines, due to its possibility to accommodate large plant variation over the flight envelope.

## 3.2 Control Lyapunov Function-Based Set Point Designs

The generalized G-B lemma approach can provide estimation of the rate of regulation. The approach assumes that engine dynamics to be the form of $\dot{x} = Ax + Bu + G(x)$, which is no loss of generality as when engine works at a certain operational condition, its dynamics can indeed by represented by the assumed form. However, it is indeed necessary to consider approaches that can be used for more general form of nonlinear expressions, e.g. $\begin{cases} \dot{x} = f(x, u) \\ y = g(x, u) \end{cases}$.

This will take us to the theory of Lyapunov. In particular, Sect. 3.2.1 gives preliminaries on Lyapunov theory; Sect. 3.2.2 initiates the control Lyapunov function method to aeroengine set point control where the numerical examples used in the above section is also used here for validation in Sect. 3.2.3. Section 3.2.4 is a diversion where it is shown that control Lyapunov function method can also be used for PID design to guarantee closed loop stabilization; finally, Sect. 3.2.5 summarizes this control Lyapunov function method.

### 3.2.1 *Preliminaries*

Lyapunov theory [38, 39] includes direct and indirect (linearized) methods, where the former is the general result upon ensuring local stability of nonlinear systems. Considering the fact that practical systems always possess certain form of (inherent) nonlinearity, the Lyapunov direct method has been the fundamental technique for nonlinear control design—control Lypunov function method. The basic idea of control Lyapunov function method is to construct a Lyapunov function (as defined below), and the stability of the system can be guaranteed by checking the directional derivative of the scalar function. The Lyapunov stability result is stated below:

**Theorem 1** *(Stability Theorem of Lyapunov): If there exists a scalar function* $V(x)$*with a continuous first order derivative within a ball* $B_{R_0}$*and:*

(1) $V(x)$*is positive definite within* $B_{R_0}$*;*
(2) $\dot{V}(x)$*is negative semi-definite within* $B_{R_0}$.

*then the equilibrium is stable; further, if the derivative* $\dot{V}(x)$*is strictly negative-definite within* $B_{R_0}$*, the equilibrium is asymptotically stable.*

**Remark** The result can be utilized for local stability analysis. If global stability is desired, the additional condition of $V(x)$ being radial-unbounded must be enforced; that is: $V(x) \to \infty \ \forall \ \|x\| \to \infty$.

### 3.2.2 Set Point Control Using Lyapunov Method

Still use the same model representation reproduced below:

$$G(x) = \begin{bmatrix} 12x_1^2 - x_2^2 \\ -1.7x_1^2 + x_2^2 \end{bmatrix} \tag{3.26}$$

$$\begin{aligned} \dot{x} &= \begin{bmatrix} -4.1476 & 1.4108 \\ 0.2975 & -3.1244 \end{bmatrix} x + G(x) + \begin{bmatrix} 0.2491 \\ 0.2336 \end{bmatrix} u \\ y &= \begin{bmatrix} 8.7379 & 0 \\ 3.3033 & 3.8052 \\ 2.1940 & 2.5749 \end{bmatrix} x \end{aligned} \tag{3.27}$$

where: input $u = \Delta W_f$, and output $y = \begin{bmatrix} PCN2R & P56/P25 & P16/P56 \end{bmatrix}^T$, with $PCN2R$ as the corrected fan speed, $P56/P25$ as the compressor pressure ratio, and $P16/P56$ as the pressure ratio between outer bypass and high-pressure turbine exit.

Now deploy a static feedback control:

$$u = K_1 x \tag{3.28}$$

where: $K_1 = \begin{bmatrix} k_1 & k_2 \end{bmatrix}$ and the nonlinear control design problem is to seek $K_1$ such that the state vector $x(t)$ is stabilized.

**Remark** In aeroengine control, static feedback control is usually utilized other than dynamic feedback for two reasons:

(1) Computational resources are limited henceforth dynamic feedback utilizing filtering or estimation is not welcomed;
(2) The states are directly associated with low-pressure and high-pressure spool speeds. For aeroengine sensors, speed sensors are most accurate henceforth whose direct feedback is desired.

Now, following the Lyapunov theory of stability, one can derive the following result.

**Theorem 2** *Control design can be guaranteed to be stable if the parameter* $K_1$ *is chosen in a bounded closure.*

***Proof*** Define the following Lyapunov function:

$$V_1(x) = x^{\mathrm{T}} P_1 x \tag{3.29}$$

where: $x = \begin{bmatrix} x_1 \ x_2 \end{bmatrix}^{\mathrm{T}}$, $P_1 = \begin{bmatrix} 1 & 0 \\ 0 & 1 \end{bmatrix}$. And the derivative of $V_1(x)$ is:

$$\dot{V}_1(x) = \dot{x}^{\mathrm{T}} P_1 x + x^{\mathrm{T}} P_1 \dot{x} \tag{3.30}$$

Insert (3.26)–(3.28) into (3.30):

$$\dot{V}_1(x) = x^{\mathrm{T}}\left(A^{\mathrm{T}} P_1 + K_1^{\mathrm{T}} B^{\mathrm{T}} P_1 + P_1 A + P_1 B K_1\right)x + G^{\mathrm{T}}(x) P_1 x + x^{\mathrm{T}} P_1 G(x) \tag{3.31}$$

Then it is known from Theorem 1 that the asymptotic stability condition at equilibrium is:

(1) $V_1(x) > 0$ and,
(2) $\dot{V}_1(x) < 0$.

Now $V_1(x) = x^{\mathrm{T}} P_1 x = x_1^2 + x_2^2 > 0$, henceforth condition (1) holds. And the objective is to choose appropriate control parameters such that (2) is satisfied.

Notice that the variables in the aeroengine model have been "corrected" implying that $x_1$ and $x_2$ are actually change of percentage in low-pressure and high-pressure rotational speeds, that is: $-1 \le x_1 \le 1$ and $-1 \le x_2 \le 1$. Insert the system equations into (3.31):

$$\begin{aligned} \dot{V}_1(x) = & -2x_1x_2^2 - 3.4x_2x_1^2 - 8.2952x_1^2 - 6.2488x_2^2 + 0.4672k_2x_2^2 + 2x_2^3 \\ & + 0.4982k_1x_1^2 + 24x_1^3 + (0.4672k_1 + 0.4982k_2 + 3.4166)x_1x_2 \end{aligned} \tag{3.32}$$

Manipulation leads to:

$$\dot{V}_1(x) = -2(1 - x_2)x_2^2 - 2(x_1 + 1)x_2^2 - 24(1 - x_1)x_1^2 - 3.4(x_2 + 1)x_1^2$$

$$+ (0.4672k_2 - 2.2488)x_2^2 + (0.4982k_1 + 19.1048)x_1^2 + (0.4672k_1 + 0.4982k_2 + 3.4166)x_1x_2 \tag{3.33}$$

Obviously, the part $-2(1-x_2)x_2^2 - 2(x_1+1)x_2^2 - 24(1-x_1)x_1^2 - 3.4(x_2+1)x_1^2 \le 0$ is always satisfied; and to make $\dot{V}_1(x) < 0$, one can dictate:

$$\begin{aligned}&(0.4672k_2 - 2.2488)x_2^2 + (0.4982k_1 + 19.1048)x_1^2\\&+(0.4672k_1 + 0.4982k_2 + 3.4166)x_1x_2 < 0\end{aligned} \tag{3.34}$$

A necessary and sufficient condition for the satisfaction of the above inequality is:

$$\begin{cases}0.4672k_2 - 2.2488 < 0\\(0.4672k_1 + 0.4982k_2 + 3.4166)^2 - 4(0.4672k_2 - 2.2488)(0.4982k_1 + 19.1048)\\< 0\end{cases} \tag{3.35}$$

Abbreviation leads to:

$$\begin{cases}k_2 < 4.8134\\(0.4672k_1 + 0.4982k_2 + 3.4166)^2 < 4(0.4672k_2 - 2.2488)(0.4982k_1 + 19.1048)\end{cases} \tag{3.36}$$

This can be illustrated in Fig. 3.7. Obviously, the system is asymptotically stable if the controller parameters are chosen in the shaded area. For example, a choice of $K_1 = \left[-12.45\ 4.81337\right]$, or $k_1 = -12.45$ and $k_2 = 4.81337$, can be checked that (3.36) is satisfied.

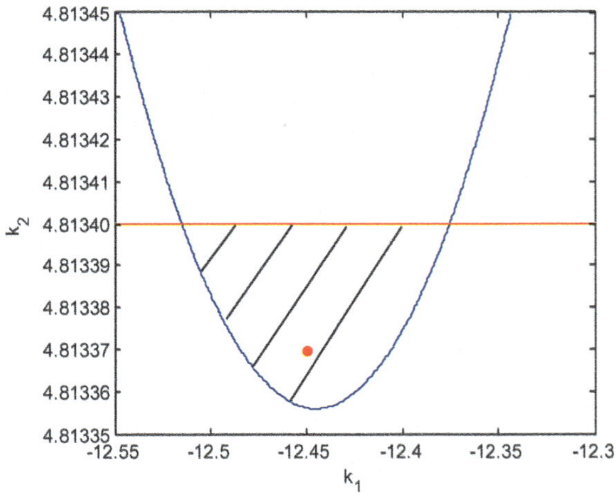


**Fig. 3.7** Feasible range for controller parameters

In summary, if the vector $K_1$ is chosen in the shaded area in Fig. 3.7, the condition for Lyapunov stability will be satisfied, and henceforth Theorem 1 can guarantee that the closed-loop system is asymptotically stable, or the state vector $x$ is stabilized.

### 3.2.3 Nonlinear Aeroengine Design Using Lyapunov Method: Numerical Example

#### Regulation Design

First consider the regulation design where $u = \Delta W_f = 0$ and the objective is to design $u = \Delta W_f = K_1 x$, such that the transient responses of both state $x$ and input $y$ can be improved.

To proceed, chose $K_1 = \left[ -12.45 \; 4.81337 \right]$ and for initial state $x_0 = \left[ 0.2 \; 0.3 \right]^T$, the transient response of $x(t)$ can be obtained as shown in Fig. 3.8, where the open-loop response is also illustrated for comparison. It is seen clearly that the closed-loop design improves the regulation performance significantly. The output response is shown in Fig. 3.9, and again the corresponding open-loop situation is also illustrated. It is seen that the system output performance is also improved with fast transition, indicating a quick rejection to disturbances.

Meanwhile, it is known from the numerical study that both the system states and outputs variables are all stabilized, should the controller parameters be chosen in the indicated feasible range. However, it is also reminded that Lyapunov stability theory is a sufficient yet not necessary condition. That is, satisfaction of Lyapunov stability theory can ensure the system stability, but those controller parameters chosen outside

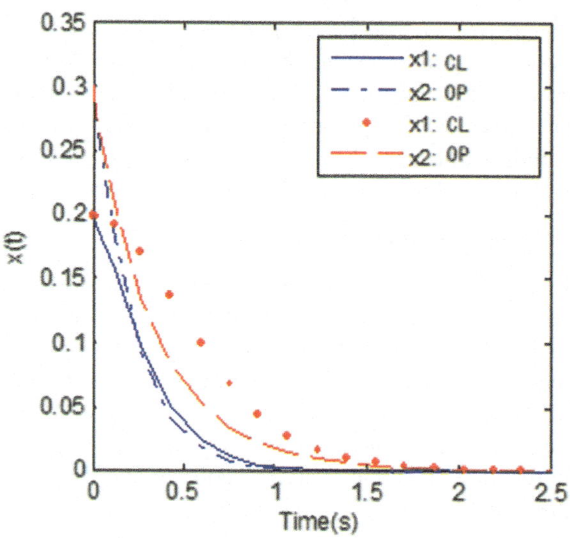


**Fig. 3.8** State response for $x_0 = [0.2 \quad 0.3]^T$: CL: Closed-loop; OP: Open-loop

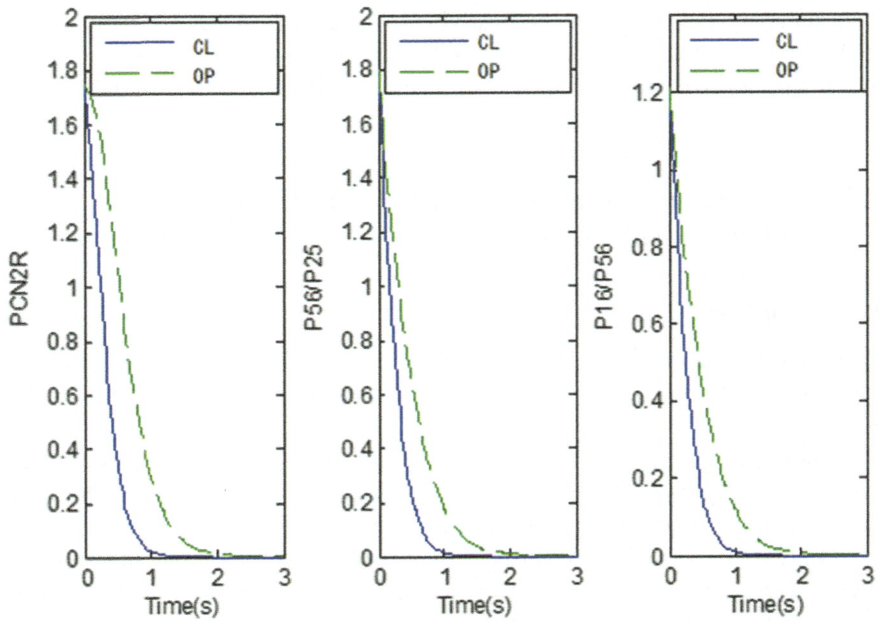


**Fig. 3.9** Output response for $x_0 = [0.2 \quad 0.3]^T$: CL: Closed-loop; OP: Open-loop

the feasible range do not necessarily cause the instability of system state or output variables. Indeed, quite often, in searching of conditions ensuring $\dot{V}_1(x) < 0$, it can be very conservative so that the feasible range for controller parameters are limited in a small area.

To validate the claim, chose the controller parameters on the boundary or even outside, e.g. $K_1 = \begin{bmatrix} -12.35 \ 4.9134 \end{bmatrix}$, then for initial state $x_0 = \begin{bmatrix} 0.2 \ 0.3 \end{bmatrix}^T$, the system responses can be obtained as shown in Figs. 3.10 and 3.11 respectively. It is seen clearly that the system is still stabilized, even though the controller parameters are chosen outside the feasible range as dictated by the Lyapunov stability theory. It is thus warned that the designer should be aware of the sufficiency or non-necessity when using Lyapunov stability theory for nonlinear aeroengine design.

***Tracking and Robustness***

Now consider the tracking performance of the above design. As is illustrated in the last section for nonlinear generalized Gronwall-Bellman lemma approach, an increment of fuel flow ratio $\Delta W_f$ indicates acceleration process of the aircraft engine. This requires fast response of $PCN2R$ to the step input of $\Delta W_f$. For certain flight height $H$ and velocity $Ma$, this implies fast response of rotation speed (of high-pressure turbine for turbofan engines). Now chose $K_1 = \begin{bmatrix} -12.45 \ 4.81337 \end{bmatrix}$, and a 10% step in fuel flow ratio, the system response is shown in Fig. 3.12. Again, for comparison, the responses of the corresponding linearized systems are also shown in the figures. A detailed discussion follows:

Fig. 3.10 State response for different controller parameters: CL: Closed-loop; OP: Open-loop

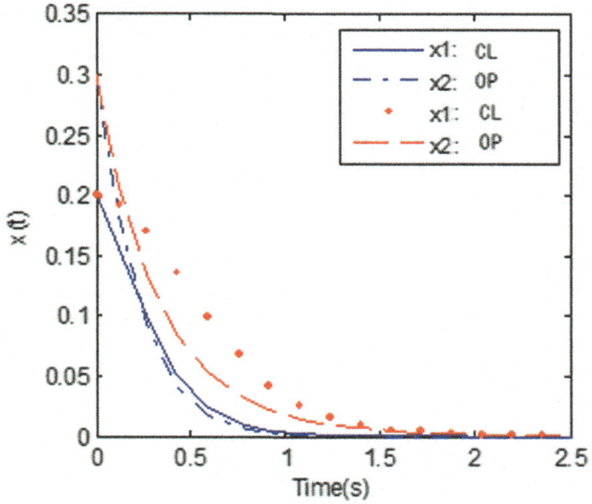


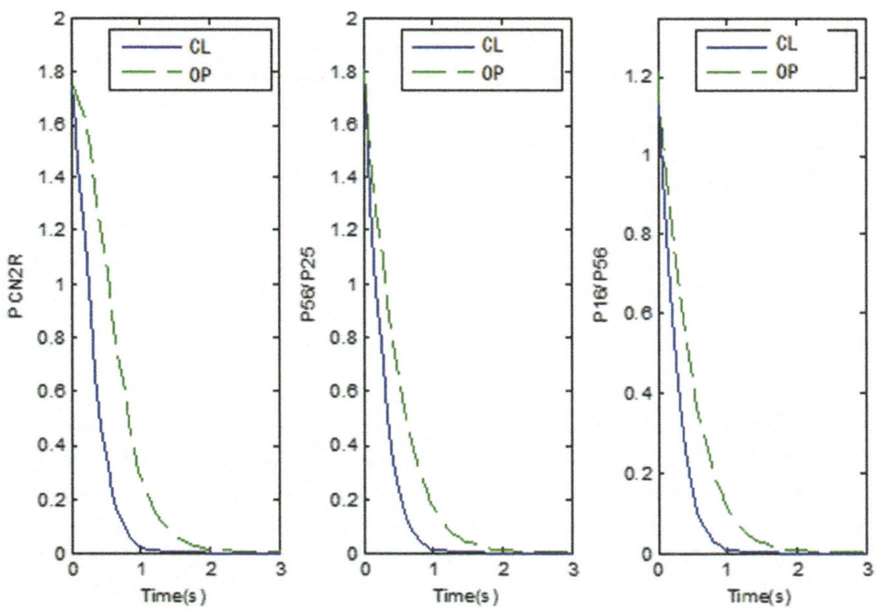


Fig. 3.11 Output response for different controller parameters: CL: Closed-loop; OP: Open-loop

(1) The result demonstrates that both linearized and nonlinear systems have improved performance under the same feedback design, indicating good tracking performance.

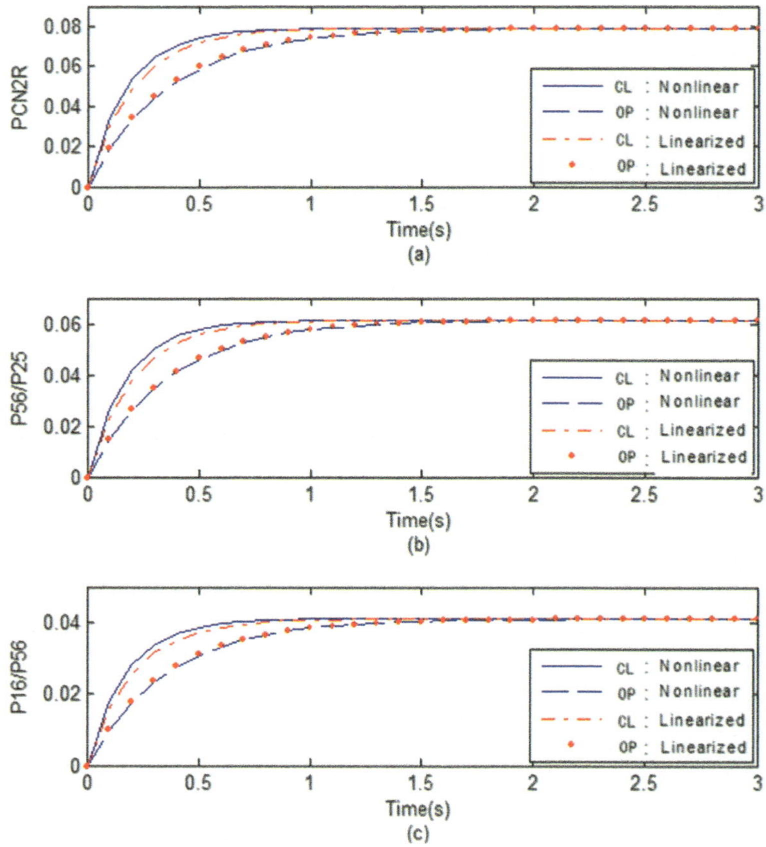


**Fig. 3.12** Output tracking performance. **a** *PCN2R*. **b** *P56/P25*. **c** *P16/P56*

(2) The comparison between the open loop and closed loop control of the linearized systems shows that the same state feedback controller also results in improved transient response, which indicates good robustness of the resulting design.

(3) The comparison between nonlinear system feedback and linear system feedback indicates that the loss of system performance is marginal. This implies that the proposed design can be applied to an extended flight envelope, which is a desirable property to be expected.

In summary, the proposed design method, namely control Lyapunov function-based set point design approach, can result in improved performance with fast transient response while possessing robustness. Numerical results validate these claims. Indeed, Lyapunov stability theory can also be used for aeroengine PID design. As is well-known, most of the aeroengine design utilizes PID control which cannot provide guarantee for resulting system stability. One idea worth of investigation is to utilize Lyapunov stability theory for PID controller parameter "detuning" in such a way that the stability will be ensured. Such an extension to PID design is discussed in the following subsection.

### 3.2.4 Set Point PID Control with Lyapunov Functions

PID control has been classical in practically all the areas of system control. As mentioned above, the current practice in aeroengine design is almost exclusively PID control. There are well-established methodology for PID control parameter tuning, but PID design is not providing closed-loop stability gaurantee but checked after the design process. Considering the benefit of Lyapunov stability theory for stability gaurantee, it is of interest to combine the two methods leading to a "stability-guaranteed PID design."

Consider certain turbo-fan engine with the following transfer function:

$$G(s) = \frac{y(s)}{u(s)} = \frac{c}{s^2 + as + b} \tag{3.37}$$

The model can then be transformed into its differential equation representation:

$$\ddot{y}(t) + a\dot{y}(t) + by(t) = cu(t) \tag{3.38}$$

Differentiation upon both sides leads to:

$$\dddot{y}(t) + a\ddot{y}(t) + b\dot{y}(t) = c\dot{u}(t) \tag{3.39}$$

The conventioal PID control design has block diagram shown in Fig. 3.13, where the PID controller has the following representation:

$$K(s) = k_p + \frac{k_i}{s} + k_d s \tag{3.40}$$

Thus, one has:

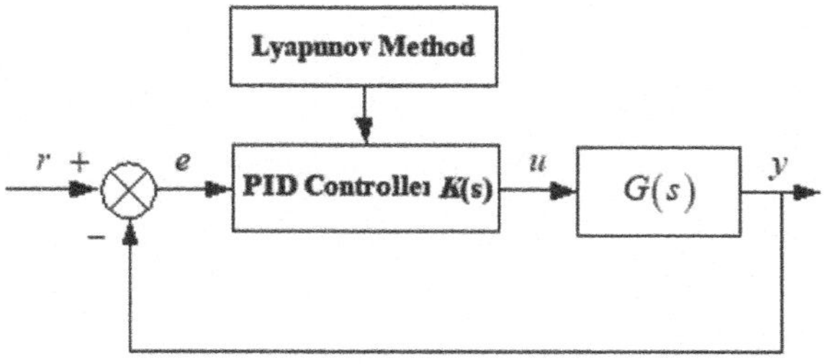


**Fig. 3.13** Set point PID control with Lyapunov functions

$$u(s) = K(s)e(s) = \left(k_p + \frac{k_i}{s} + k_d s\right)e(s) \tag{3.41}$$

A Laplace inverse transform to (3.41) results in:

$$u(t) = k_i \int e(t)dt + k_p e(t) + k_d \frac{d}{dt} e(t) \tag{3.42}$$

where: $u(t)$ is the input; $e(t)$ is the error signal defined as $e(t) = r(t) - y(t)$, with $r(t)$ the reference and $y(t)$ the output. Then one has:

$$\begin{cases} \dot{e}(t) = \frac{d(r-y)}{dt} = -\dot{y}(t) \\ \ddot{e}(t) = \frac{d(\dot{e}(t))}{dt} = -\ddot{y}(t) \\ \dddot{e}(t) = \frac{d(\ddot{e}(t))}{dt} = -\dddot{y}(t) \end{cases} \tag{3.43}$$

This can be represented in a state-space form as:

$$\dot{x}(t) = Ax(t) + B\dot{u}(t) \tag{3.44}$$

where: the state variable is $x(t) = \begin{pmatrix} e(t) \\ \dot{e}(t) \\ \ddot{e}(t) \end{pmatrix}$. And the matrix coefficients can be obtained as:

$$A = \begin{bmatrix} 0 & 1 & 0 \\ 0 & 0 & 1 \\ 0 & -b & -a \end{bmatrix}, \quad B = \begin{bmatrix} 0 \\ 0 \\ -c \end{bmatrix} \tag{3.45}$$

Now choose a state feedback control:

$$\dot{u}(t) = K_2 x(t) \tag{3.46}$$

where: $K_2 = \begin{bmatrix} k_1 & k_2 & k_3 \end{bmatrix}$. Then one can obtain:

$$\dot{u}(t) = \begin{bmatrix} k_1 & k_2 & k_3 \end{bmatrix} \begin{bmatrix} e(t) \\ \dot{e}(t) \\ \ddot{e}(t) \end{bmatrix} = k_1 e(t) + k_2 \dot{e}(t) + k_3 \ddot{e}(t) \tag{3.47}$$

Integration upon both sides of (3.47) leads to:

$$u(t) = k_1 \int e(t)dt + k_2 e(t) + k_3 \frac{d}{dt} e(t) \tag{3.48}$$

Then one can establish the following "correspondence" of parameters with PID control: $\left[\, k_i \; k_p \; k_d \,\right] = \left[\, k_1 \; k_2 \; k_3 \,\right]$.

Now to take advantage of Lyapunov stability theory, one chooses the following Lyapunov function:

$$V_2(x) = x^{\mathrm{T}} P_2 x \tag{3.49}$$

Taking first-order derivative results in:

$$\dot{V}_2(x) = \dot{x}^{\mathrm{T}} P_2 x + x^{\mathrm{T}} P_2 \dot{x} \tag{3.50}$$

Inserting (3.44), (3.46) into (3.50):

$$\dot{V}_2(x) = x^T \left( A^T P_2 + K_2^T B^T P_2 + P_2 A + P_2 B K_2 \right) x \tag{3.51}$$

Now consider a specific example where the aeroengine operates on ground state with the following transfer function [40]:

$$G(s) = \frac{y(s)}{u(s)} = \frac{1.139}{s^2 + 2.493s} \tag{3.52}$$

where: system output y is the low-pressure rotational speed $n_1$(r/min); system input u is the aeroengine fuel flow $W_{\mathrm{f}}$(kg/h). It is then ready to derive the matrices:

$$A = \begin{bmatrix} 0 & 1 & 0 \\ 0 & 0 & 1 \\ 0 & 0 & -2.493 \end{bmatrix}, \quad B = \begin{bmatrix} 0 \\ 0 \\ -1.139 \end{bmatrix} \tag{3.53}$$

Again one needs to ensure (1) $V_2(x) > 0$; and (2) $\dot{V}_2(x) < 0$. Now for $V_2(x) = x^{\mathrm{T}} P_2 x$, take $P_2 = \begin{bmatrix} 6 & 3 & 0.6 \\ 3 & 321 & 11 \\ 0.6 & 11 & 15 \end{bmatrix}$, and the eigenvalues of $P_2$ are:$\lambda_1 = 5.943$; $\lambda_2 = 14.633$; and $\lambda_3 = 321.424$. Thus $P_2$ is positve definite, and henceforth $V_2(x) > 0$. That is, condition (1) now holds. To ensure condition (2) satisfed $\dot{V}_2(x) < 0$, one can aim to satisfy the following inequality:

$$A^T P_2 + K_2^T B^T P_2 + P_2 A + P_2 B K_2 < 0 \tag{3.54}$$

This is a typical linear matrix inequality with $A$, $B$, $P_2$ being known and $K_2$ to be determined. In Matlab, the toolbox for Linear Matrix Inequality (LMI) is available where solvers can be readily used for obtaining the unknown parameters. In this case, it is the $K_2$ as the to-be-determined parameters. An application of LMI toolbox results in $K_2 = \left[\, 0.087 \; 8.847 \; 2.769 \,\right]$, and thus the system stability is ensured with the following PID controller parameters: $k_i = 0.087$; $k_p = 8.847$; $k_d = 2.769$. For

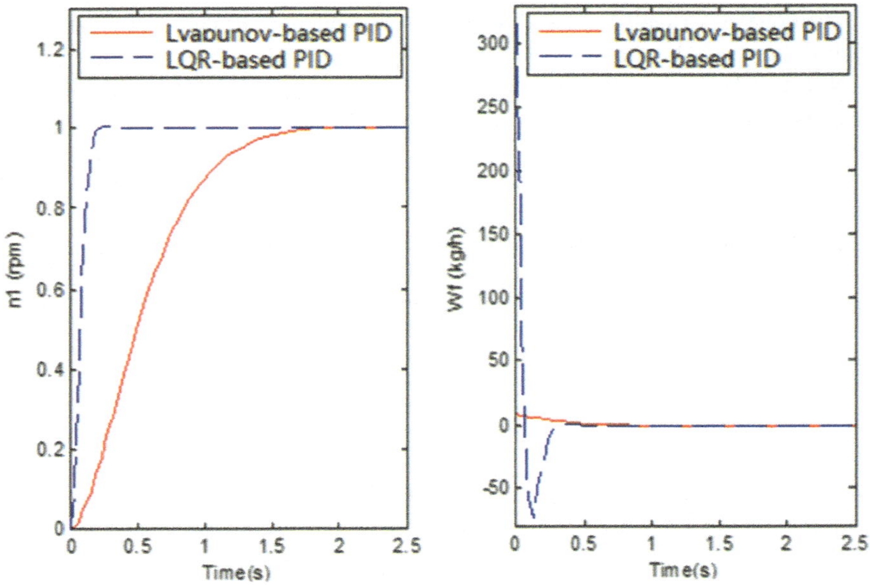


(a) (b)

**Fig. 3.14** Step response without control constraint. **a** Rotational speed. **b** Fuel flow

comparison, we will taken the design from Ref. [40] where an LQR-based approach is proposed with the following PID controller parameters: $k_i = 1.000$; $k_p = 316.303$; $k_d = 21.480$.

When there is no constraint on control, the step response is shown in Fig. 3.14, where the fuel flow is also provided. It is seen that although the transient response of Lyapunov-based PID design is not as fast as that of LQR-based one, the proposed design does provide no overshoot, no steady-state error with a regulation time 1.8s, satisfying the aeroengine control system requirement.

In addition, it is noticed that the LQR-based PID design has a "jump" in fuel flow injection, indicating the fuel metering devices will experience additional payload, which is not desirable as well as economic. This is in stark contrary to the Lyapunov-based PID design.

Now enforce a control constraint within $\pm 8.85$, or the fuel flow is restricted with the values of the fuel consumption for Lyapunov-based approach, the result is shown in Fig. 3.15. It is seen clearly that Lyapunov-based design provides good performance in rotational speed response while the fuel consumption smoothly decreases to zero. While LQR-based design is influenced significantly with a seriously degraded performance. It results in an overshot of 10% which is unacceptable, not to mention that fuel flow is severely distorted leading to wear even damage to fuel metering devices. This demonstrates clearly the superiority of the proposed Lyapunov-based PID design or PID detuning.

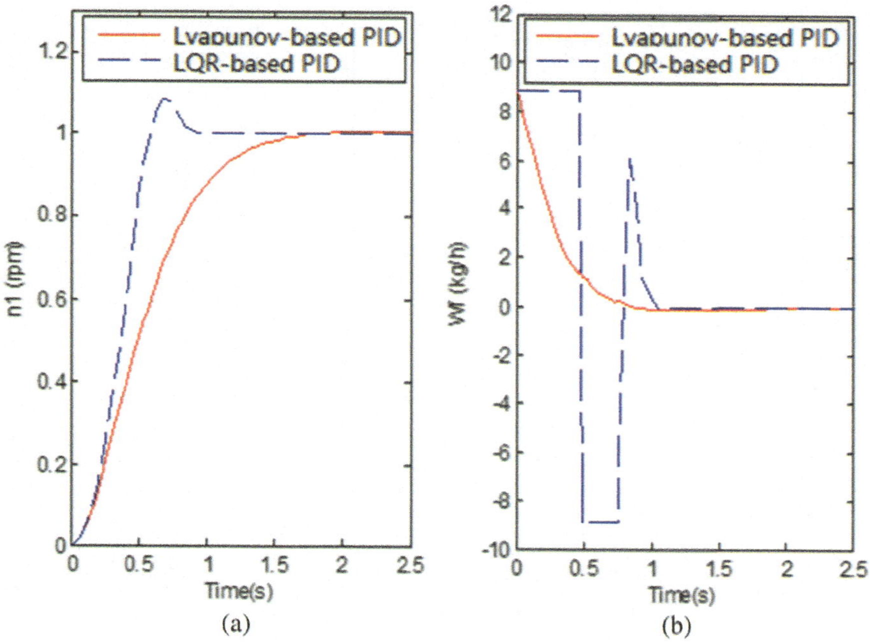


**Fig. 3.15** Step response with control constraint ±8.85. **a** Rotational speed. **b** Fuel flow

For a "fair" comparison, a control constraint within ±3 is enforced, or the fuel flow is restricted below the values of both Lyapunov-based and LQR-based approaches, the result is shown in Fig. 3.16. It is seen that Lyapunov-based design still provides good performance in both output and input; while the LQR-based design results in totally unacceptable system performance, e.g. slow response, large overshoot, and unsmooth fuel flow consumption.

Thus, it is validated that Lyapunov-based method can ensure the stability of the PID design, accomplishing a detuning of PID parameters. Although PID parameters are obtained through extensive experimentation and henceforth very costly thus not expected to change during the validation process, the proposal does provide a simple, sound and easy design procedure for PID detuning, e.g. parameters are detuned when aeroengine operates in different environmental conditions.

### *3.2.5 Summary for Lyapunov-based Design*

Control Lyapunov function-based method for aeroengine set point control is introduced and numerical study is provided for validation of the proposed approach. It is demonstrated that Lyapunov-based design can ensure system stability while providing desired system performance. Meanwhile, the method can also be used to

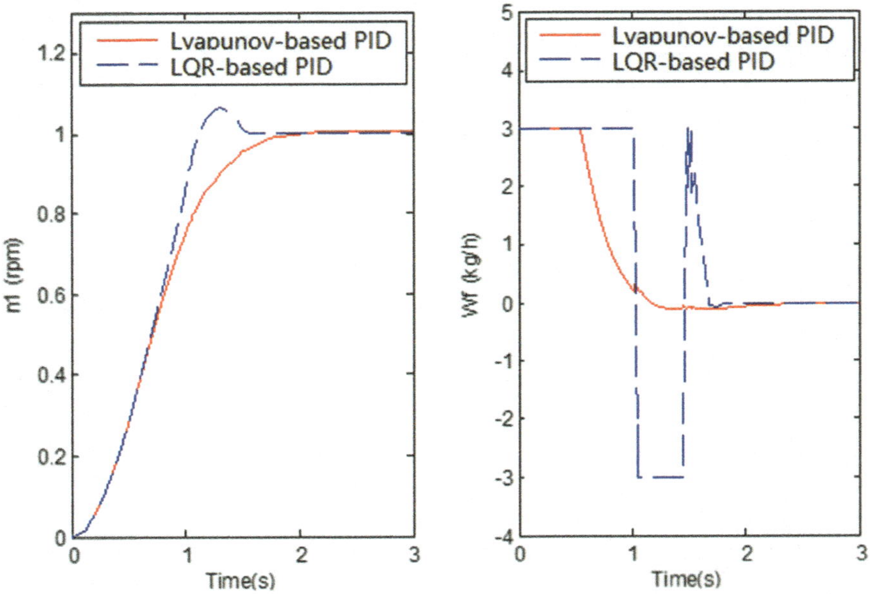


(a) (b)

**Fig. 3.16** Step response with control constraint ±3. **a** Rotational speed. **b** Fuel flow

PID detuning. A comparison with LQR-based PID design shows the advantage of Lyapunov-based PID design for parameter detuning.

## 3.3 Joint Design with Generalized GB Lemma and Lyapunov Function

Lyapunov function-based design can provide stability guaranteed yet it is realized that the important issue of convergence rate is not given consideration. While the generalized Gronwall-Bellman lemma-based approaches introduced in Sect. 3.1 can be used for bound estimation, it is thus interesting to integrate both methods. The basic idea is to ensure system stability and thus using Lyapunov-based method to obtain controller parameters. The resulting controller parameters are henceforth restricted within a feasible area; and within this area, the generalized GB lemma method can be used for fast convergence [41].

Now following the example in the last section, it is assumed that the initial state $x_0 = \left[-0.2\ -0.3\right]^T$, then $\|x_0\| = 0.3606 < 0.3627$, the system transient response $x(t)$ is illustrated in Fig. 3.17. It is seen that regulation time is 1.2 s, and the system performance is improved with a fast transient without overshoot.

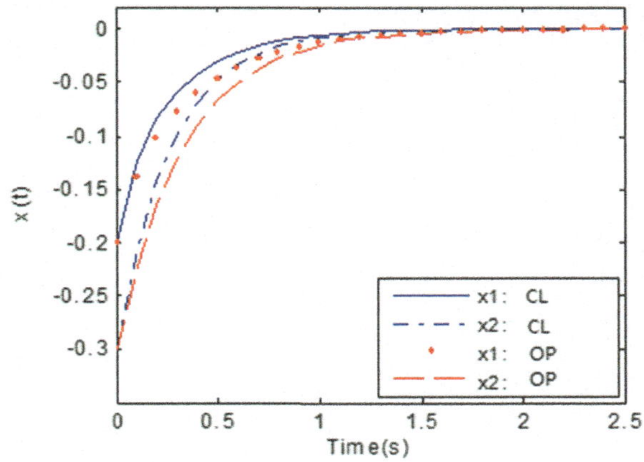


**Fig. 3.17** System response for $x_0 = [-0.2 \quad -0.3]^T$

Now from the inequality $\|x(t)\| \leq \frac{\|x_0\|e^{-4.3524t}}{(1-2.7571\|x_0\|)}$, the bounding behavior $\|x(t)\|$ can be calculated for $x_0 = \begin{bmatrix} -0.2 & -0.3 \end{bmatrix}^T$, as shown in Fig. 3.18. Thus it is known that when the initial condition satisfies $\|x_0\| < 0.3627$, the norm of the system states will be bounded by $\|x(t)\| \leq \left(\|x_0\|e^{-4.3524t}\right)/(1 - 2.7571\|x_0\|)$. From Fig. 3.18, it is seen that $\|x(t)\|$ is strictly bounded within the theoretical boundary thus the convergence can be pre-estimated with the integration of generalized GB lemma.

Yet it is also noticed from Fig. 3.18 that $\|x(t)\|$ is tightly bounded with the theoretical boundary. This indicates that the requirement upon the initial condition $\|x_0\|$ is too restricted. Desired performance can still be expected, should the range on $\|x_0\|$ be enlarged. Now take $x_0 = \begin{bmatrix} -0.3 & -0.5 \end{bmatrix}^T$, it is known $\|x_0\| = 0.5831 > 0.3627$, then for the same controller, the transient response $x(t)$ is illustrated in Fig. 3.19. The regulation time is 1.4 s without overshoot. The performance is thus satisfactory, confirming with the above analysis.

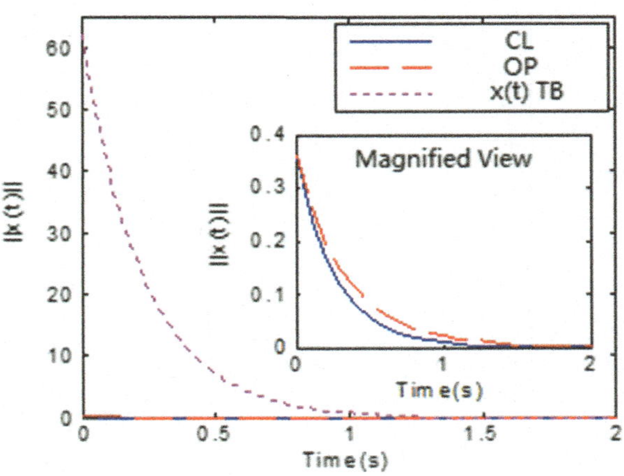


**Fig. 3.18** Convergence of $\|x(t)\|$ with theoretical boundary: CL-closed loop; OP: open loop; TB-theoretical boundary

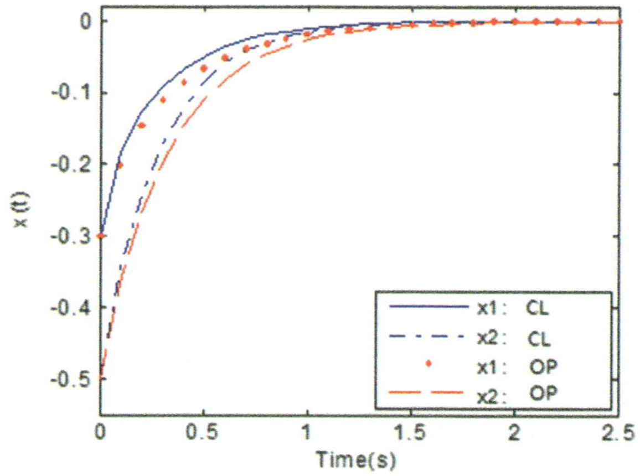


**Fig. 3.19** System response for $x_0 = [-0.3 \quad -0.5]^T$

Now the output $y = \left[\,PCN2R\ P56/P25\ P16/P56\,\right]^T$ for initial condition $x_0 = \left[\,-0.2\ -0.3\,\right]^T$ can be plotted as in Fig. 3.20. It is seen that the system performance is improved over the open-loop situation.

To further testify the capability for noise suppression, random noises with 5, 10, 15, and 20% of the input magnitude are injected. For a clear comparison, the noise is injected with different magnitude while comparing with the situation without noise. Firstly, 5 and 10% noise levels are considered and the results are shown in Fig. 3.21, where it is seen that the output response performance is good when 5% noise level is injected; while there exist small fluctuations, the performance is still accepted for 10% noise level.

The situation with 15 and 20% noise levels is shown in Fig. 3.22, where it is seen that the output response starts obvious fluctuations when 15% noise level is injected; while the performance becomes so deteriorated into non-acceptance should the noise

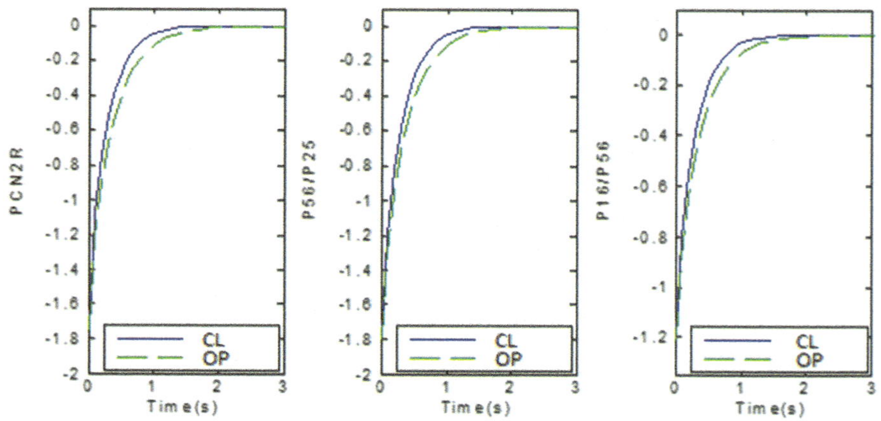


**Fig. 3.20** Output response for $x_0 = [-0.2 \quad -0.3]^T$

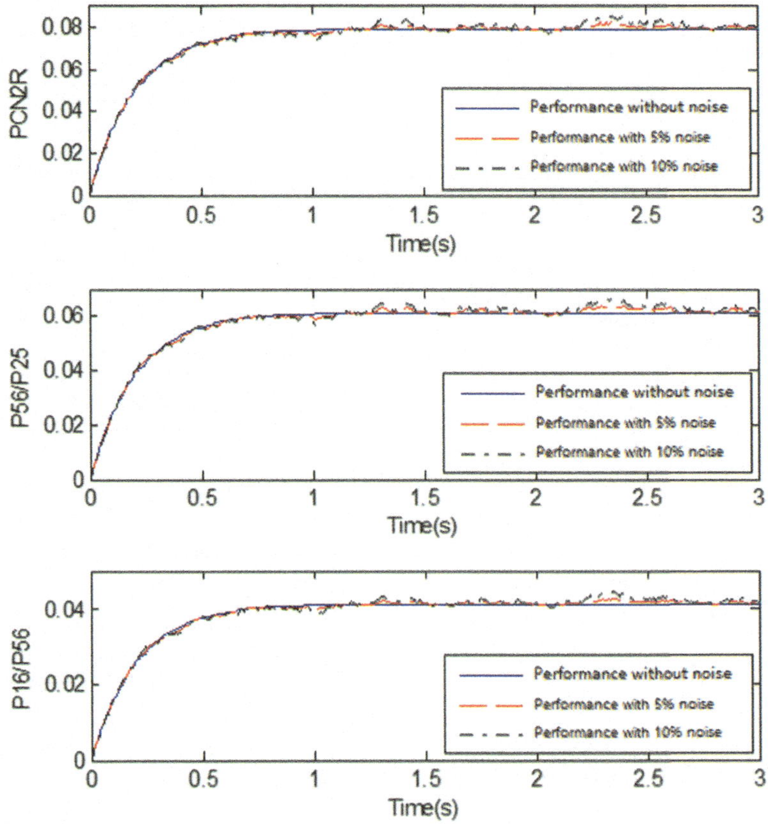


**Fig. 3.21** Output performance with 5 and 10% noise level

level be increased to 20% level. Consequently, the system can resist noise level within 10%; and a further increase of noise level will lead to decreased system performance, henceforth degenerated robustness of control design.

On the other hand, control design as discussed is model based, indicating that the model can accurately capture the aeroengine dynamics. However, due to the large envelope, the aero-thermal dynamics varies significantly while possessing strong nonlinearity and time-variability. Although some simplified nonlinear model can be used, the model for control design is usually too simplied to be able to have enough confidence. Thus, model uncertainty must be accounted for during control design processes. Model uncertainty is caused from many factors, including modeling error, model variations, model parameter floating due to aging etc. Thus, control design must be robust against the model uncertainty.

To further validate the joint design, the matrices $A, B, C, D$ are perturbed. A simple scenario is considered here where matrix $A$ has a variation along the diagonal elements of $\pm 10$ and $\pm 20\%$. When $+10\%$ is considered, one has:

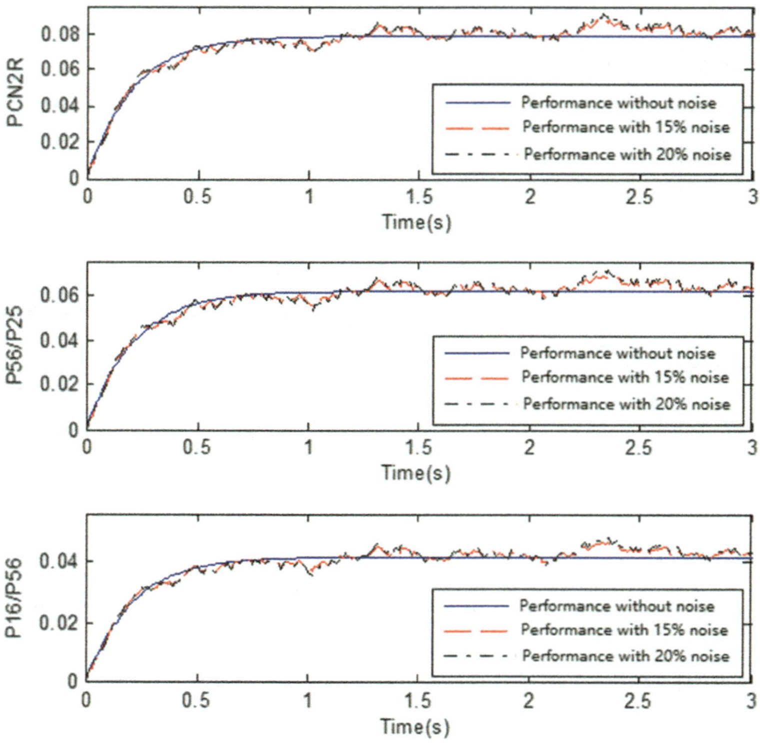


**Fig. 3.22** Output performance with 15 and 20% noise level

$$A = \begin{bmatrix} 1.1(-4.1476) & 1.4108 \\ 0.2975 & 1.1(-3.1244) \end{bmatrix};$$

while a $-10\%$ is considered, one has:

$$A = \begin{bmatrix} 0.9(-4.1476) & 1.4108 \\ 0.2975 & 0.9(-3.1244) \end{bmatrix};$$

similarly, for $+20\%$ case,

$$A = \begin{bmatrix} 1.2(-4.1476) & 1.4108 \\ 0.2975 & 1.2(-3.1244) \end{bmatrix},$$

and for $-20\%$ case, one can readily calculated to have

$$A = \begin{bmatrix} 0.8(-4.1476) & 1.4108 \\ 0.2975 & 0.8(-3.1244) \end{bmatrix}.$$

Figure 3.23 shows the output response for a variation of ±10% in matrix $A$. The comparison with the nominal performance shows that the system is stable when a ±10% model variation is present; indeed, the loss of control performance is marginal. Now should a ±20% model variation is considered, it is seen from Fig. 3.24 that the system can retain stability; even the transient response is improved for +20% variation, with a regulation time 1 s without overshot; for −20% variation, the transient response slows a bit but the overall performance is retained.

In summary, even a ±20% model uncertainty occurs, the loss of system performance is acceptable; and the control performance can almost retain as the nominal performance. This "diversion" shows that the proposed design also possesses robustness against model uncertainty, which is a requirement for any control design proposal.

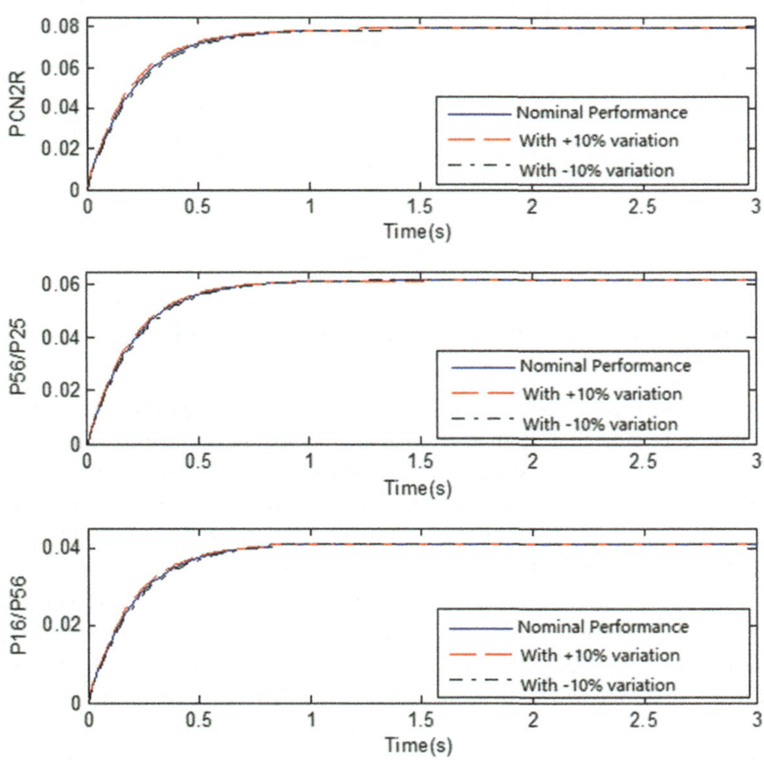


**Fig. 3.23** Output performance with 10% model uncertainty

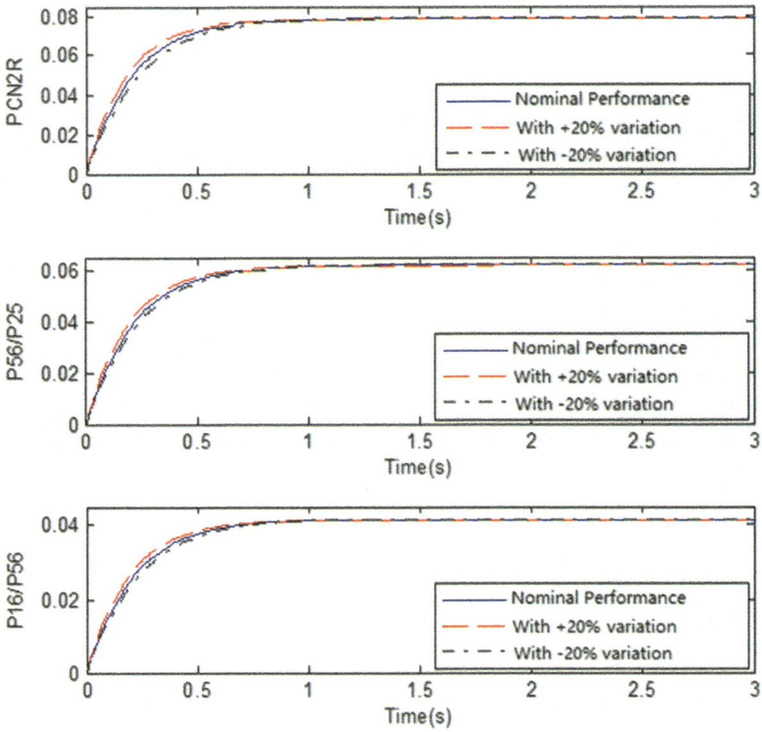


**Fig. 3.24** Output performance with 20% model uncertainty

## 3.4 Restricted Control Lyapunov Function Method

For both generalized Gronwall-Bellman Lemma and Lyapunov function based approaches, it is seen that control feedback $u = Kx$ required full state feedback, and in cases where the states are not all available, observers or filters can be used for estimation of the states. However, for current state of technology, it is still not possible to have an accurate estimation using model-based observers and filters. This is so due to the fact that nonlinear estimation and filtering theories relies on accurate model or extensive computations. And it is unfortunate that both aspects are not feasible in FADEC system. Therefore, it is of practical interest to consider using less sensors or actuators to achieve control purposes. This section considers the extreme case of using only one sensor and one actuator for achieving set point control.

This single sensor and single actuator approach to nonlinear control of engines has many important implications that have been overlooked in current design. First of all, single sensor and single actuator design is desirable since actuators are usually expensive and heavy and hence should be avoided in design. Therefore, it is very useful to use single actuator and single sensor to control multivariate performance variables. Secondly, there is a practical need to use a single sensor and single actuator, e.g. in engine control systems, the rotational speeds of the shafts for low-pressure compressor and high-pressure compressor cannot be controlled simultaneously due to the aero-dynamical couplings. Finally, it can be prohibitively expensive for system implementation other than the restricted sensing and actuation control. Therefore, the single sensor and single actuator approach represents an important research direction to be investigated.

From control system perspective, single sensor and single actuator can be thought of as a static output feedback control problem. Static output feedback design is not only theoretically important, but also practically appealing due to its simplicity and reliability for implementation. Indeed, any dynamic output feedback problem can be transformed into a static output feedback problem [42]. However, the problem of static output feedback control is challenging and still remains open, even for linear systems. This famous open problem in linear control theory can be stated as finding a static gain matrix $K$ such that $u = Ky$ stabilizes the linear system $\dot{x} = Ax + Bu$, $y = Cx$, e.g. all the eigenvalues of the matrix $(A + BKC)$ have strictly negative real parts. This deceptively simple problem is in fact complicated (even *NP*-hard) as minimization of spectral radius is known to be neither convex nor locally Lipchitz [43, 44]. As a consequence, efforts are devoted to sufficient conditions for stability and the corresponding efficient algorithms for realization, e.g. q-SNM method for control with [45] and without constraints [46]; gradient sampling/non-smooth optimization technique [47]; multi-objective optimization method [48] etc.

A literature survey shows that static output feedback control for nonlinear systems is not as widely investigated as its linear counterpart (see the recent paper [49]). For nonlinear systems, the non-convexity issues persist, and no general result is available. However there are still advances in both leading to sufficient conditions and developing efficient algorithms for specific problems. For example, the static output feedback stabilization problem is examined by an iterative sums of squares approach in [49], based on a semi-definite programming (SDP) method to solve state-dependent LMIs [50–52]. In [53] and [54], the problem is converted into the solvability of Hamilton–Jacobi equation, enabling approximation methods to obtain solutions (hence inherently suboptimal, see [55] and references therein for techniques to solve the Hamilton–Jacobi equation); a computational scheme of solving the static output feedback control problem for a class of polynomial nonlinear systems is proposed in [56], also exploring the sum of squares decomposition-based methodology.

In this section, however, a Lyapunov function-based method is introduced for the design of static output feedback stabilization. The contributions can be summarized as follows: (1) it is different from the existing optimization-based approaches, hence not computationally demanding; (2) control action is restricted to one sensor and one actuator, henceforth to be the simplest possible implementation scheme, while all the results so far need more than one sensor to be implementable; (3) numerical studies are provided for control of aircraft engines, where the single sensor and single actuator approach is pretty novel. To understand comprehensively the significance of the second contribution, it is pointing out that this paper is partially motivated by the author's effort to controlling multiple performance variables using only restricted controls. This is a "remote" control problem where only one sensor signal is available to one local actuator, but performance is also required at remote locations. This point will be further delineated in the next section.

This section is structured as follows: Sect. 3.4.1 formulates the problem to be studied before presenting the main results in Sect. 3.4.2. While the results are applicable for disturbance-free case, control design with bounded disturbances is addressed in Sect. 3.4.3. Section 3.4.4 provides a numerical study for validation of the proposed design, and some important issues associated with the design methodology are also discussed. Finally, Sect. 3.4.5 concludes the section.

### 3.4.1 Problem Formulation

Consider a general nonlinear system $\dot{x} = f(x)$ that is rewritten into the following form:

$$\dot{z}_1 = g_1(z_1, z_2), \dot{z}_2 = g_2(z_1, z_2) \tag{3.55}$$

where: the state variable $z_1$ is a scalar; the state variable $z_2 = (x_1 \cdots x_{N-1})^T$ is a vector, with the subsystem $\dot{z}_2 = g_2(z_1, z_2)$ being Lipchitz in a neighbourhood of $z_1 = 0$ while $\dot{z}_2 = g_2(0, z_2)$ being uniformly exponentially stable about $z_2 = 0\ \forall z$. The output equation is $y = h(z_1, z_2)$ with $h$ being nonlinear function of the states.

The objective is to stabilize the system (3.55) using only static and single input:

$$u = ky \tag{3.56}$$

where $k$ is a constant.

That is to say, the system should be stabilized using a single sensor and a single actuator. As stated above, this is an important problem that is often overlooked in practical designs since, on one hand, actuators are expensive and/or heavy and are hence avoided in a system design [57]; on the other hand, there are many situations where in-service information from $z_2$ is simply either not obtainable or prohibitively expensive to do so. As a consequence, control design can only proceed with $z_1$ (often remote to $z_2$) and its feedback action, but with the performance objective of controlling both $z_1$ and $z_2$, refer to Fig. 3.25.

Such examples are abounding, particularly in the inter-connected large-scale structures. For example, in the helicopter or submarine rotor blade control, only the shaft acceleration is available for feedback, while both the shaft vibration and the blade vibration should be attenuated. Obviously here the difficulty is that it is not practically viable to permanently locate sensors into the blades or prohibitively expensively to do so. Although there exist approaches such as integrating smart

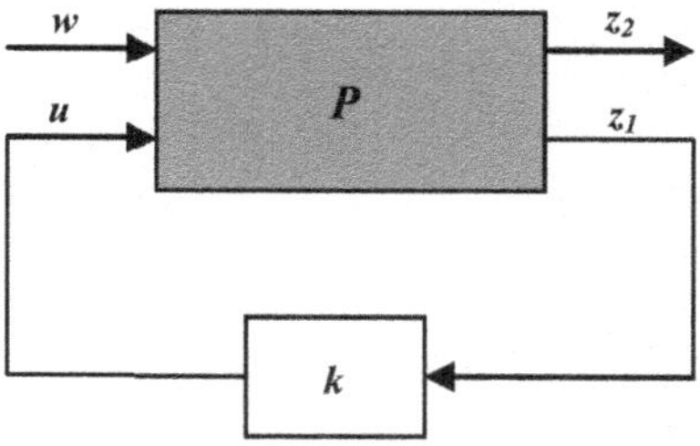


**Fig. 3.25** Static feedback control using single sensor and single actuator. This is actually a remote control problem [58]

materials into the blades, these solutions are costly and unproven in real operational environments. The research presented here should be considered as an ideal alternative solution to the remote control problem [58].

### 3.4.2 Main Results on Restricted Control Design

To state the main result, first notice that the assumption that the subsystem $\dot{z}_2 = g_2(z_1, z_2)$ is Lipchitz in a neighbourhood of $z_1 = 0$ implies, there can be found a positive constant $\lambda_1$ such that:

$$\|g_2(z_1, z_2) - g_2(0, z_2)\| \leq \lambda_1 \|z_1\| \tag{3.57}$$

Also $\forall z$, the assumption that $\dot{z}_2 = g_2(0, z_2)$ is uniformly exponentially stable about $z_2 = 0$ implies there exist a Lyapunov function $V_0(z_2)$ and two positive numbers $\lambda_2, \lambda_3$ such that:

$$\dot{V}_0(z_2) = \frac{\partial V_0(z_2)}{\partial z_2} g_2(0, z_2) \leq -\lambda_2 \|z_2\|^2 \tag{3.58}$$

$$\left\| \frac{\partial V_0(z_2)}{\partial z_2} \right\| \leq \lambda_3 \|z_2\| \tag{3.59}$$

Furthermore, $g_1(z_1, z_2)$ being Lipchitz implies there can be found another positive number $\lambda_4$ such that:

$$\|g_1(z_1, z_2)\| \leq \lambda_4 \|z_1\| \tag{3.60}$$

The main results of the paper can now be stated:

**Theorem 1** *Consider the following single-input nonlinear control system:*

$$\dot{z}_1 = g_1(z_1, z_2) + u_1, \quad \dot{z}_2 = g_2(z_1, z_2) \tag{3.61}$$

*If the following condition holds:*

$$z_1 g_1(z_1, z_2) + z_1 k(z_1, z_2) + \frac{\partial V_0(z_2)}{\partial z_2} g_2(z_1, z_2) \leq 0 \tag{3.62}$$

*where $k(z_1, z_2)$is a polynomial in $z_1$and $z_2$, then the single-input and static state feedback control*

$$\begin{aligned} &u_1 = k(z_1, z_2) - \tfrac{cV(z_1, z_2)}{z_1},\ V(z_1, z_2) = \tfrac{1}{2}z_1^2 + V_0(z_2) \text{ if } z_1 \neq 0 \\ &u_1 = 0 \qquad\qquad\qquad\qquad\qquad\qquad\qquad\qquad \text{if } z_1 = 0 \end{aligned} \tag{3.63}$$

*exponentially stabilizes the closed loop system.*

***Proof*** Consider the Lyapunov function candidate $V(z_1, z_2) = \frac{1}{2}z_1^2 + V_0(z_2)$. Differentiating it along the trajectory of (6) leads to:

$$\begin{aligned} \dot{V}(z_1, z_2) &= z_1 \dot{z}_1 + \frac{\partial V_0(z_2)}{\partial z_2} g_2(z_1, z_2) \\ &= z_1 \left[ g_1(z_1, z_2) + k(z_1, z_2) - \frac{cV(z_1, z_2)}{z_1} \right] + \frac{\partial V_0(z_2)}{\partial z_2} g_2(z_1, z_2) \\ &= z_1 g_1(z_1, z_2) + z_1 k(z_1, z_2) - cV(z_1, z_2) + \frac{\partial V_0(z_2)}{\partial z_2} g_2(z_1, z_2) \\ &\leq -cV(z_1, z_2) \end{aligned}$$

Hence the nonlinear control system (3.61) is guaranteed to be exponentially stabilized. (END).

**Theorem 2** *For the nonlinear control system* (3.61), *the single input and static state feedback control*

$$\begin{aligned} &u_1 = k_1 z_1 - \tfrac{cV(z_1, z_2)}{z_1} \quad \text{if } z_1 \neq 0 \\ &u_1 = 0 \qquad\qquad\qquad\quad \text{if } z_1 = 0 \end{aligned} \tag{3.64}$$

*exponentially stabilizes the closed loop system, provided the static gain $k_1$satisfies the following condition:*

$$k_1 \leq -\lambda_4 - \frac{\lambda_1^2 \lambda_3^2}{4\lambda_2} \tag{3.65}$$

***Proof*** In condition (3.62), substitute $k(z_1, z_2) = k_1 z_1$, the following is obtained:

$$z_1 g_1(z_1, z_2) + k_1 \|z_1\|^2 + \frac{\partial V_0(z_2)}{\partial z_2} g_2(z_1, z_2) \leq 0 \tag{3.66}$$

Or:

$$z_1 g_1(z_1, z_2) + k_1\|z_1\|^2 + \frac{\partial V_0(z_2)}{\partial z_2}[g_2(z_1, z_2) - g_2(0, z_2)] + \frac{\partial V_0(z_2)}{\partial z_2} g_2(0, z_2) \le 0 \tag{3.67}$$

Hence, from the conditions (3.57)–(3.60):

$$z_1 g_1(z_1, z_2) \le \lambda_4\|z_1\|^2, \ \frac{\partial V_0(z_2)}{\partial z_2}[g_2(z_1, z_2) - g_2(0, z_2)] \le \lambda_1\lambda_3\|z_1\|\,\|z_2\|$$

Therefore, the condition (3.67) will be satisfied if the following inequality stands:

$$\lambda_4\|z_1\|^2 + k_1\|z_1\|^2 + \lambda_1\lambda_3\|z_1\|\,\|z_2\| - \lambda_2\|z_2\|^2 \le 0 \tag{3.68}$$

For any $z_1$ and $z_2$, condition (3.68) will hold if $\lambda_1^2\lambda_3^2 + 4\lambda_2(k_1 + \lambda_4) \le 0$ and this gives the condition (3.65).

To recap, if the static gain $k_1$ is such chosen that satisfies condition (3.65) $k_1 \le -\lambda_4 - \frac{\lambda_1^2\lambda_3^2}{4\lambda_2}$, then the Lyapunov function candidate $V(z_1, z_2) = \frac{1}{2}z_1^2 + V_0(z_2)$ along the trajectory of the nonlinear control system (3.61) satisfies $\dot{V}(t) \le -cV(t)$. Thus, exponential stabilization is achieved via the single input, static state feedback control (3.64). (END).

**Remark 1** Comparing with the controller (3.63), the controller (3.64) uses only a static gain $k(z_1, z_2) = k_1 z_1$, thus simplifies control implementation. The static gain $k_1$ is only constrained by the condition (3.65).

However, the controller (3.64) is still "complex" as its implementation requires sensing of both $z_1$ and $z_2$. As is pointed out above, there are many situations where in-service information from $z_2$ is either not obtainable or cost-effective. It is therefore desirable to design a controller based only on $z_1$, whiling controlling both $z_1$ and $z_2$. This important problem is solved in the following result.

**Theorem 3** *For the nonlinear control system* (3.61),*if the static gain $k_1$satisfies*

$$k_1 \le -\lambda_4 - \frac{\lambda_1^2\lambda_3^2}{4\lambda_2} \tag{3.65}$$

*then the single input, single state feedback control*

$$\begin{array}{ll} u_1 = k_1 z_1 - \frac{c|z_1|}{z_1} & \text{if } z_1 \ne 0 \\ u_1 = 0 & \text{if } z_1 = 0 \end{array} \tag{3.69}$$

*asymptotically stabilizes the closed loop system.*

***Proof*** Consider again the Lyapunov function candidate $V(z_1, z_2) = \frac{1}{2}z_1^2 + V_0(z_2)$ and differentiate it along the trajectory of (3.60):

$$\begin{aligned}
\dot{V}(z_1, z_2) &= z_1\dot{z}_1 + \frac{\partial V_0(z_2)}{\partial z_2} g_2(z_1, z_2) \\
&= z_1\left[g_1(z_1, z_2) + k_1 z_1 - \frac{c|z_1|}{z_1}\right] + \frac{\partial V_0(z_2)}{\partial z_2} g_2(z_1, z_2) \\
&= z_1 g_1(z_1, z_2) + k_1\|z_1\|^2 + \frac{\partial V_0(z_2)}{\partial z_2}[g_2(z_1, z_2) - g_2(0, z_2)] \\
&\quad + \frac{\partial V_0(z_2)}{\partial z_2} g_2(0, z_2) - c|z_1| \\
&\leq \lambda_4\|z_1\|^2 + k_1\|z_1\|^2 + \lambda_1\lambda_3\|z_1\|\|z_2\| - \lambda_2\|z_2\|^2 - c|z_1|
\end{aligned}$$

Therefore, from the proof of Theorem 2, the condition $k_1 \leq -\lambda_4 - \frac{\lambda_1^2\lambda_3^2}{4\lambda_2}$ implies that the inequality $\lambda_4\|z_1\|^2 + k_1\|z_1\|^2 + \lambda_1\lambda_3\|z_1\|\|z_2\| - \lambda_2\|z_2\|^2 \leq 0$ always holds. Hence $\dot{V}(z_1, z_2) \leq -c|z_1| < 0\ \forall z_1 \neq 0$. This proves the result. □

**Remark 2** It is remarkable that the design parameter $k_1$ in controller (3.64) and controller (3.69) satisfies the same inequality. However, comparing with controller (3.64), controller (3.69) requires only information from $z_1$, hence is the single sensor and single actuator approach to controlling the nonlinear system. For static output feedback control problem, this implies that the output equation becomes $y = h(z_1, z_2) = z_1 + \frac{c}{k_1 z_1}$. This is not really a handicap as most of the iterative sums of squares approaches for nonlinear static output feedback control only admit linear output equation $y = Cx$ ($C$ is constant) to relieve the computational demand, see the numerical examples in [49].

### *3.4.3 Restricted Design with Bounded Disturbance*

Any realistic system consists of disturbances and/or noises in one form or another, and consequently the corresponding control design must possess certain robustness property. This section considers the problem of static feedback control of nonlinear systems with bounded disturbance. Consider the following single-input control system:

$$\dot{z}_1 = g_1(z_1, z_2) + u_1 + d_1(t), \dot{z}_2 = g_2(z_1, z_2) + d_2(t) \tag{3.70}$$

where: $d_1(t)$, $d_2(t)$ are disturbances entering the two subsystems, satisfying $\|d_1(t)\| \leq l_1$ and $\|d_2(t)\| \leq l_2$, respectively. The problem is to stabilize the system (3.70) using a single sensor and a single actuator. This problem is addressed in the following important result:

**Theorem 4** *For the nonlinear control system* (3.70),*if the static gain $k_1$satisfies*

$$k_1 \le -\lambda_4 - \frac{\lambda_1^2\lambda_3^2}{4\lambda_2} \tag{3.65}$$

*then the single input and static state feedback control*

$$\begin{aligned} u_1 &= k_1 z_1 - \tfrac{cV(z_1,z_2)}{z_1} \text{ if } z_1 \ne 0 \\ u_1 &= 0 \qquad\qquad\qquad \text{ if } z_1 = 0 \end{aligned} \tag{3.64}$$

*exponentially stabilizes the closed loop system to the region $Q$ defined by:*

$$\begin{aligned} Q = \{&z_1 \ge 0, z_1 \ge 0, z_2 \ge 0, \\ &(\lambda_4 + k_1)\|z_1\|^2 + \lambda_1\lambda_3\|z_1\|\|z_2\| - \lambda_2\|z_2\|^2 + l_2\lambda_3\|z_2\| + l_1\|z_1\| = 0\} \end{aligned}$$

*Furthermore, for the same static gain $k_1$satisfying* (3.65),*the single input, single state feedback control*

$$\begin{aligned} u_1 &= k_1 z_1 - \tfrac{c|z_1|}{z_1} \text{ if } z_1 \ne 0 \\ u_1 &= 0 \qquad\qquad\quad \text{ if } z_1 = 0 \end{aligned} \tag{3.69}$$

*asymptotically stabilizes the closed loop system to $Q$.*

***Proof*** Consider the Lyapunov function candidate $V(z_1, z_2) = \frac{1}{2}z_1^2 + V_0(z_2)$ and differentiate it along the trajectory of (3.70):

$$\begin{aligned} \dot{V}(z_1, z_2) &= z_1\dot{z}_1 + \frac{\partial V_0(z_2)}{\partial z_2}[g_2(z_1, z_2) + d_2(t)] \\ &= z_1\left[g_1(z_1, z_2) + k_1 z_1 - \frac{cV(z_1, z_2)}{z_1} + d_1(t)\right] \\ &+ \frac{\partial V_0(z_2)}{\partial z_2}[g_2(z_1, z_2) + d_2(t)] \\ &= z_1 g_1(z_1, z_2) + k_1\|z_1\|^2 + \frac{\partial V_0(z_2)}{\partial z_2}g_2(z_1, z_2) + z_1 d_1(t) \\ &+ \frac{\partial V_0(z_2)}{\partial z_2}d_2(t) - cV(z_1, z_2) \\ &\le (\lambda_4 + k_1)\|z_1\|^2 + \lambda_1\lambda_3\|z_1\|\|z_2\| - \lambda_2\|z_2\|^2 \\ &+ l_2\lambda_3\|z_2\| + l_1\|z_1\| - cV(z_1, z_2) \end{aligned}$$

Consider $F(z_1, z_2) \equiv (\lambda_4 + k_1)\|z_1\|^2 + \lambda_1\lambda_3\|z_1\|\|z_2\| - \lambda_2\|z_2\|^2 + l_2\lambda_3\|z_2\| + l_1\|z_1\|$. After a change of coordinates, $F(z_1, z_2)$ can be rewritten into:

$$\begin{aligned} F(z_1, z_2) = (\lambda_4 + k_1)(\|z_1\| - \|z_{10}\|)^2 + \lambda_1\lambda_3(\|z_1\| - \|z_{10}\|)(\|z_2\| - \|z_{20}\|) \\ - \lambda_2(\|z_2\| - \|z_{20}\|)^2 \end{aligned}$$

$$-(\lambda_4 + k_1)\|z_{10}\|^2 + \lambda_2\|z_{20}\|^2 - \lambda_1\lambda_3\|z_{10}\|\|z_{20}\| \tag{3.71}$$

where $\|z_{10}\|$ and $\|z_{20}\|$ are determined by the method of undetermined coefficients:

$$\|z_{10}\| = -\frac{2l_1\lambda_2 + l_2\lambda_1\lambda_3^2}{4\lambda_2\lambda_4 + 4k_1\lambda_2 + \lambda_1^2\lambda_3^2} \tag{3.72–1}$$

$$\|z_{20}\| = \frac{2l_2\lambda_3(k_1 + \lambda_4) - l_1\lambda_1\lambda_3}{4\lambda_2\lambda_4 + 4k_1\lambda_2 + \lambda_1^2\lambda_3^2} \tag{3.72–2}$$

Substitute into (17) and manipulate, it is seen that $F(z_1, z_2) = 0$ is defined by the equation:

$$\begin{aligned}&(\lambda_4 + k_1)(\|z_1\| - \|z_{10}\|)^2 + \lambda_1\lambda_3(\|z_1\| - \|z_{10}\|)(\|z_2\| - \|z_{20}\|)\\&\quad - \lambda_2(\|z_2\| - \|z_{20}\|)^2 = -C\end{aligned} \tag{3.73}$$

where: $C = \frac{1}{4\lambda_2}\left[l_2^2\lambda_3^2 - \frac{(2l_1\lambda_2 + l_2\lambda_1\lambda_3^2)^2}{4\lambda_2\lambda_4 + 4k_1\lambda_2 + \lambda_1^2\lambda_3^2}\right]$.

Now let $Q = \left\{ (z_1,\ z_2,\ F(z_1, z_2)) \middle|\ z_1 \geq 0,\ z_2 \geq 0,\ F(z_1, z_2) = 0 \right\}$, it is seen that for a specific design $k_1$, $Q$ encloses a domain on $z_1 - z_2$ plane (Fig. 3.26) whose

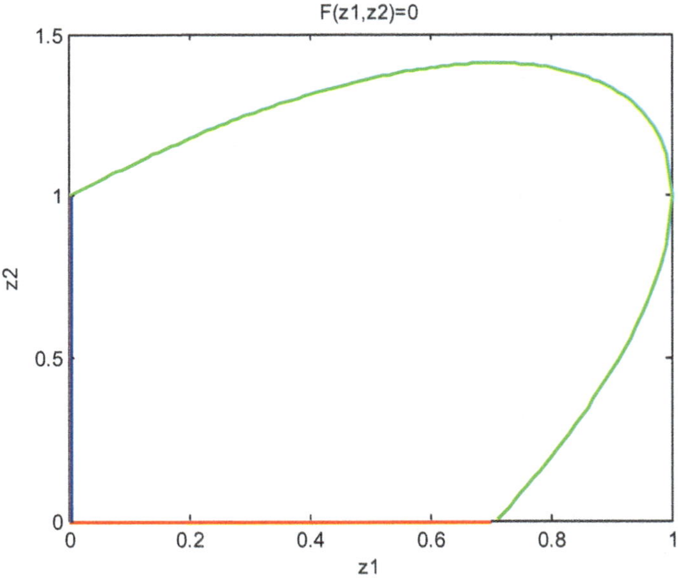


**Fig. 3.26** An example of domain $Q$

area depends on $C$, henceforth on the disturbance bounds $l_1$ and $l_2$, e.g. $l_1 = l_2 = 0$, $Q$ reduces to a single point at the origin (0, 0, 0).

If now the static control gain is such chosen satisfying:

$$\lambda_1^2\lambda_3^2 + 4\lambda_2(k_1 + \lambda_4) \le 0 \tag{3.74}$$

For any $z_1$ and $z_2$, there will have $F(z_1, z_2) \le 0$ outside $Q$ on $z_1 - z_2$ plane with equality obtained on the boundary. That is for $(z_1, z_2) \in \Re^2 - Q$:

$$\dot{V}(z_1, z_2) \le F(z_1, z_2) - cV(z_1, z_2) \le -cV(z_1, z_2) \tag{3.75}$$

The system will be exponentially stabilized to $Q$ via the single input, static state feedback control (3.64). A similar argument can be used to show that the single input, single state feedback control (3.69) can asymptotically stabilize the system (3.70) to $Q$. (END).

**Remark 3** Comparing with the disturbance free case, the static gain $k_1$ can only provide stability to a region $Q$ other than the origin. However it is noted that the enclosed area depends on the disturbance bounds $l_1$ and $l_2$, e.g. $l_1 = l_2 = 0$, $Q$ reduces to the origin (0, 0, 0); on the other hand, for given $l_1$ and $l_2$, a sufficiently small $k_1$ will reduce $Q$, which can be demonstrated from the generalized Gronwall-Bellman lemma but nevertheless provide a guidance for design.

### *3.4.4 Nonlinear Set Point Control of Aeroengines: Restricted Design*

This subsection presents a realistic study on nonlinear control of aircraft engines. First consider the following nonlinear model around certain operating point without disturbance entering in:

$$\begin{aligned} \Delta\dot{n}_1 &= -4.1476\Delta n_1 + 1.4108\Delta n_2 + 12\Delta n_1^2 - \Delta n_2^2 + \Delta w_f \\ \Delta\dot{n}_2 &= 0.2975\Delta n_1 - 3.1244\Delta n_2 - 1.7\Delta n_1^2 + \Delta n_2^2 \end{aligned} \tag{3.76}$$

where: $\Delta n_1$ represents the rotational speed change in low pressure turbines while $\Delta n_2$ represents the rotational speed change in high pressure turbines; $\Delta w_f$ is the fuel flow ratio as control input. The model is adopted from the Intelligent Engine Control project [59]. The nonlinear part has been augmented for the investigation of advanced control concepts, comparing with linear controls exclusively utilized in the state-of-art design.

Now let $z_1 = \Delta n_1$, $z_2 = \Delta n_2$, it is seen that the subsystem $\dot{z}_2 = g_2(z_1, z_2)$ is smooth in a neighbourhood of $z_1 = 0$ and $\dot{z}_2 = g_2(0, z_2)$ is uniformly exponentially stable about $z_2 = 0\ \forall z$. The assumption made in this paper is satisfied. Take again $V(z_1, z_2) = \frac{1}{2}z_1^2 + V_0(z_2) = \frac{1}{2}(z_1^2 + z_2^2)$. From Theorem 1, it is known that the controller (3.64) can exponentially stabilize the system (3.76) if the polynomial $k(z_1, z_2)$ is such chosen that the following condition is satisfied:

$$z_1 g_1(z_1, z_2) + z_1 k(z_1, z_2) + z_2 g_2(z_1, z_2) \leq 0 \tag{3.77}$$

Now choose the static gain $k(z_1, z_2) = k_1 z_1$, and substitute (3.76) into (3.77) leads to:

$$(k_1 - 4.1476 - 1.7z_2 + 12z_1)z_1^2 + 1.7083z_1 z_2 + (z_2 - z_1 - 3.1244) \leq 0 \tag{3.78}$$

By the technique of completing the squares, together with the consideration of $|\Delta n_1| \leq 1$, $|\Delta n_2| \leq 1$, it is obtained that inequality (3.78) will hold if the static gain $k_1$ satisfies:

$$k_1 < -10.2073 \tag{3.79}$$

Take $k_1 = -11$ and the single-input and static state feedback control is given by:

$$u_1 = -11z_1 - \frac{cV(z_1, z_2)}{z_1} \tag{3.80}$$

Theorems 1, 2 and the above analysis guarantee that controller (3.80) will asymptotically stabilize the engine (3.76). A simulation with initial condition $\left[\, z_1(0)\ z_2(0) \,\right] = \left[\, -0.5\ 1.0 \,\right]$, $c = 2$ confirms the assertion, see Figs. 3.27 and 3.28, also shown is the performance of the PI control for benchmarking.

The single sensor and single actuator solution can now be obtained from Theorem 3, which asserts that for the same control gain $k_1$, the following controller can provide asymptotical stability:

$$\begin{aligned} u_1 &= -11z_1 - \tfrac{c|z_1|}{z_1} && \text{if } z_1 \neq 0 \\ u_1 &= 0 && \text{if } z_1 = 0 \end{aligned} \tag{3.81}$$

It is again observed that the asymptotic stabilizer leads to a relatively sluggish response than the exponential one, but the control effort is reduced significantly. See Figs. 3.29 and 3.30.

Now consider the case where disturbances affect both $\Delta n_1$ and $\Delta n_2$ with $d_1(t) = 2.5\sin(60t) + 0.05U(0, 1)$ and $d_2(t) = 1.5\sin(80t) + 0.1U(0, 1)$. To be more general, both disturbances consist of a random noise superposed by a low frequency signal, representing typical pulsation dynamics of the engine fuel flow regulator. According to Theorem 4, for the same static gain $k_1$, the single input, single state feedback control (3.81) will asymptotically stabilize the closed loop system to a region Q as shown in Figs. 3.31 and 3.32. It is seen that the proposed design provides much better performance than that of PI control. Even remarkable is

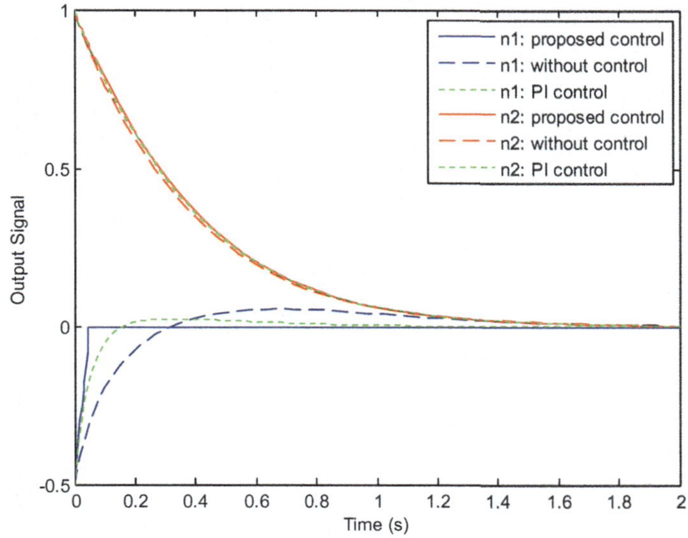


Fig. 3.27 Regulation of aeroengine via single input and static state feedback

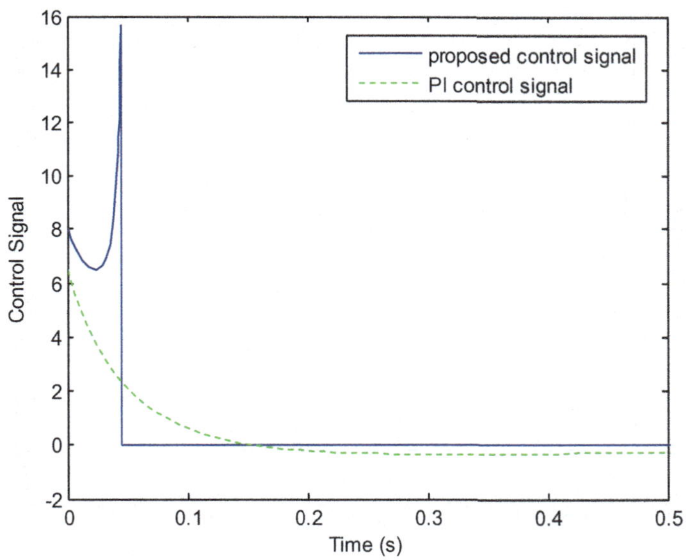


Fig. 3.28 Response of control signal $u_1$ in (3.80)

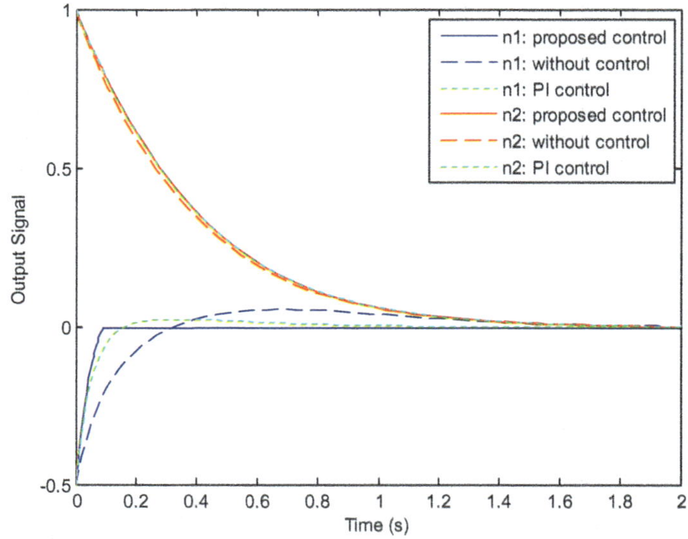


**Fig. 3.29** Regulation aeroengine via single input and single actuator without disturbance

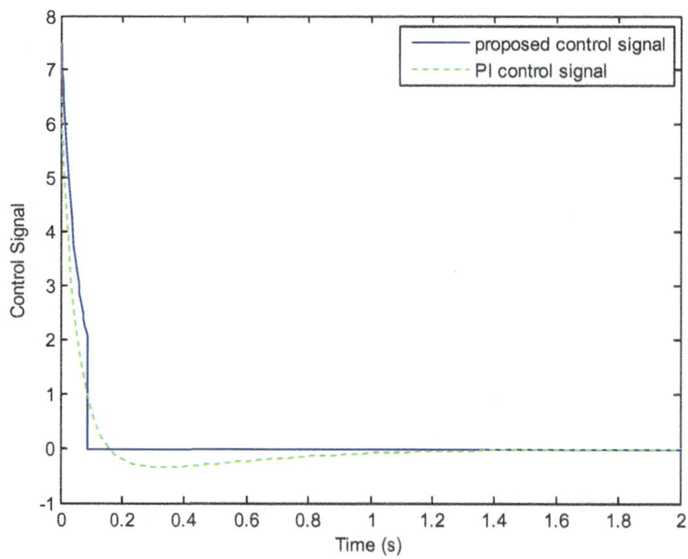


**Fig. 3.30** Response of control signal $u_1$ in (3.81) and PI control for disturbance-free case

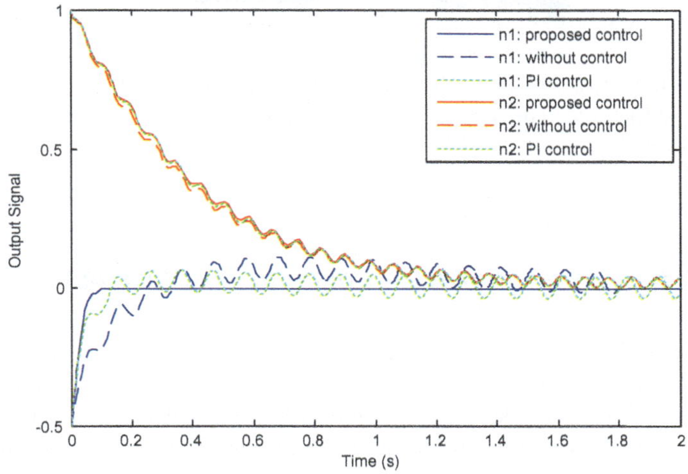


**Fig. 3.31** Regulation of aeroengine via single input and single actuator with disturbances

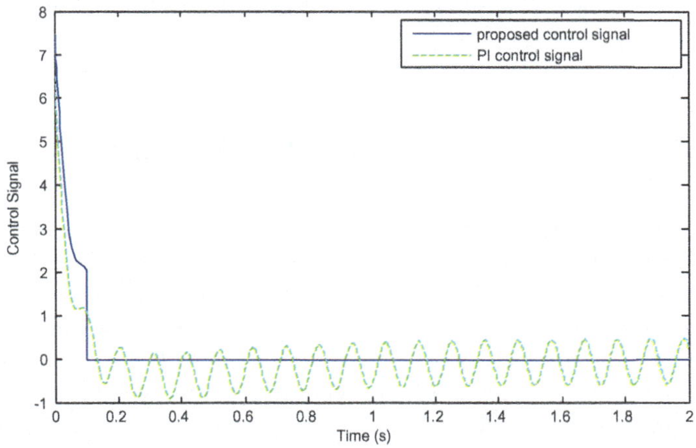


**Fig. 3.32** Response of control signal $u_1$ in (27) and PI control for disturbance case

that almost asymptotical stability to origin for $\Delta n_1$ is achieved in spite of disturbances while this is not the case for PI control that responds violently to the disturbances.

**Remark 4** Simulations show that the regulation performance of $\Delta n_1$ is much better than that of $\Delta n_2$. This is so since the low-pressure turbine has only aerodynamic coupling with the high-pressure turbine. However, it is exactly due to this aerodynamic coupling that prevents the direct control of $\Delta n_1$ and $\Delta n_2$ simultaneously, necessitating the single actuator approach to the control of aircraft engines.

**Remark 5** The state of the art in aircraft engine control utilizes linear design methodologies exclusively. Advanced nonlinear control can accommodate large plant variation over the flight envelope, hence representing one of the important research directions in the field. The results presented here contribute to this important area of research as well.

### *3.4.5 Summary for Restricted Lyapunov-based Design*

The problem of single sensor and single actuator stabilization of nonlinear systems has been discussed. Some sufficient conditions have been obtained and these theoretical results have been validated through the nonlinear regulation of aircraft engines. Different from the existing optimization-based approaches, the proposed design method is relatively easy and non-computationally demanding. Even important is the single sensor and single actuator approach for nonlinear control that can be very desirable for practical engineering systems. In fact, the method proposed here solves the challenging remote-control problem that has been encountered in diverse areas of applications.

## 3.5 Finite Time Set Point Control of Aeroengines

The above design approach introduced in the last section can be further refined to provide fast regulation of engines. This takes advantage of finite time control where the system can be regulated within a finite time. Indeed, the approaches up to now are utilizing a continuous control concept where settling time for the engine rotational speed can take a long time (theoretically, it takes infinitely long to converge into the steady state value). Recently the concept of finite time stability and finite time control has been proposed where regulation is achieved in a (predetermined) finite time, see the references [60–66]. This is very suitable for turbofan engine set point controller design and it is thus desirable to apply finite time control methodology to engine control. However, one important thing that has been overlooked both in the engine control community and in the theory of finite time control is the issue of *non-simultaneous controllability for low-pressure shaft speed and high –pressure shaft speed*. This can be explained by considering a state space representation of engine dynamics at certain operation condition:

$$\begin{bmatrix} \Delta\dot{n}_L \\ \Delta\dot{n}_H \end{bmatrix} = \begin{bmatrix} g_1(\Delta n_L, \Delta n_H) \\ g_2(\Delta n_L, \Delta n_H) \end{bmatrix} + \begin{bmatrix} b_1 \\ b_2 \end{bmatrix} \Delta W_f \tag{3.82}$$

where $n_H$ and $n_L$ denote the rotational speeds of the high-pressure compressor and fan representatively; $\Delta$ represents the deviation of engine variables from their steady state values for the specified operation condition, and thus $\Delta W_f$ is the fuel flow ratio as the input; $g_1(\Delta n_L, \Delta n_H)$ and $g_2(\Delta n_L, \Delta n_H)$ are the corresponding dynamics; $B \equiv \begin{bmatrix} b_1 \\ b_2 \end{bmatrix}$ is the input matrix. The fact that the rows of input matrix B are not zero, that is both $b1$ and $b2$ are not zero, implies that the fuel flow ratio $\Delta W_f$ can control $n_H$ and $n_L$ simultaneously. In practice, however, the dynamics of high-pressure compressor/high-pressure turbine system is aero-dynamically coupled with that of fan/low-pressure turbine system. Therefore it is not possible to control the engine through controlling $n_H$ and $n_L$ simultaneously. This does represent a challenging task for control design of this system, although the controlled variables can be either $n_H$ or $n_L$ depending on the control strategy as well as different engine vendors.

This section attempts to resolve the above challenging problem with a finite time control setup. It is structured as follows: the concept of finite time stability is introduced and finite time control problem is formulated in Sect. 3.5.1; Sect. 3.5.2 aims to establish a series of important results on finite time regulation of two-spool turbofan engines with one spool speed control; Sect. 3.5.3 considers the robustness issue while a numerical example is then provided to validate the proposed design in Sect. 3.5.4. Finally, conclusions and further issues are discussed in Sect. 3.5.5.

### *3.5.1 Preliminary on Finite Time Control and Problem Formulation*

We first introduce the concept of finite time stability, and explain its difference from conventional continuous time exponential stability, together with its implication to turbofan engine set point regulation.

**Definition** For the nonlinear system $\dot{x} = f(x)$ where $x \in R^N$ and $f : R^N \to R^N$ is a smooth nonlinear vector function with $x_e$ as an equilibrium, if there exists a constant $T > 0$, such that $\lim_{t \to T} \|x(t)\| = 0$ and $\|x(t)\| \equiv 0$ for $t \geq T$, then the stabilization of the nonlinear system is achieved in a finite time.

From the definition, it is seen that finite time stability demands $\|x(t)\| \equiv 0$ after time $T$, and $T$ is a constant that is strictly less than infinity, while conventional continuous time stability only achieves this for $T \to \infty$, e.g. for step response, settling time is defined to be the time attracted within 5 or 2% confidence level of steady state value, otherwise it takes infinitely long to converge to the steady state. It is in this sense that conventional continuous time exponential stability can be called infinite time stability.

**Remark 1** It is noted that there are other methods to achieve finite time stability. The early finite time control is open-loop such as minimum-energy control [67]. The lack of robustness of open-loop control motives closed-loop methods for finite time control. And it is found in [68, 69] that finite time stability can be achieved by modifying the performance index of optimal control into non-quadratic for double integrator systems. Another closed-loop method is through the augmentation of sliding mode control, namely terminal sliding mode [70, 71]. However, both closed-loop methods are based on non-continuous state feedback and it is well-known that chattering even instability may occur. The search for closed-loop, continuous state feedback, and finite time stability finally motives the finite time control methodology introduced here.

The fundamental result on finite time stability is provided by the following proposition:

**Proposition 1** *The nonlinear system* $\dot{x} = f(x)$*can be stabilized in a finite time if there exists a continuous and positive-definite function* $V(x)$*that satisfies* $\dot{V}(x) \leq -cV^{\eta}(x), \forall t \geq t_0$*and* $V(x_0) \geq 0$, *where* $c > 0$*and* $0 < \eta < 1$*are two constants,* $t_0$*and* $x_0$*are initial conditions. Then* $\|x(t)\| \equiv 0$*for* $t \geq T$*with* $T$*given by* $T = t_0 + \frac{V^{1-\eta}(t_0)}{c(1-\eta)}$.

**Remark 2** Turbofan engine set-point controller is required to provide fast regulation of shaft speed. Conventional design approaches such as backstepping or feedback linearization achieves perfect regulation with infinite time theoretically. Finite time control introduced here can regulate the shaft speed within time duration of $T = t_0 + \frac{V^{1-\eta}(t_0)}{c(1-\eta)}$.

### *3.5.2 Finite Time Set Point Designs*

With the preliminaries presented in the last section, it is now embarking on the finite time control design for regulating a two-spool turbofan engine with only one spool speed control. A two-spool engine working on the fuel flow control mode is represented in Eq. (3.82) replicated here for easy reference:

$$\begin{bmatrix} \Delta\dot{n}_L \\ \Delta\dot{n}_H \end{bmatrix} = \begin{bmatrix} g_1(\Delta n_L, \Delta n_H) \\ g_2(\Delta n_L, \Delta n_H) \end{bmatrix} + \begin{bmatrix} b_1 \\ b_2 \end{bmatrix} \Delta W_f \tag{3.82}$$

As commented above, conventionally it is assumed that both $b1$ and $b2$ are not zero, implying that the fuel flow ratio input can control both the low-pressure spool speed and high-pressure spool speed *simultaneously*. In practice, however, only one of the spool speeds can be controlled directly. Henceforth either $b1$ or $b2$ must be zero.

**Remark 3** Which spool speed is to be controlled depends on different control modes that the engine is working on. They are thus different for different engine vendors and even different over the flight envelope for the same engine operation, e.g. during the starting process, high-pressure spool speed is controlled while during cruise the control authority is switched to the low-pressure spool speed control mode by FADEC (full authority digital engine control).

Without any loss of generality, it is assumed that $n_L$ is to be controlled and model (3.82) can be written in the following form:

$$\begin{bmatrix} \Delta \dot{n}_L \\ \Delta \dot{n}_H \end{bmatrix} = \begin{bmatrix} g_1(\Delta n_L, \Delta n_H) \\ g_2(\Delta n_L, \Delta n_H) \end{bmatrix} + \begin{bmatrix} \Delta W_f \\ 0 \end{bmatrix} \tag{3.83}$$

where the coefficient $b_1$ is suppressed within the fuel flow ratio. Therefore, the set point control problem is to design a control law $\Delta W_f = h(\Delta n_L, \Delta n_H)$, such that both the states $\Delta n_L$ and $\Delta n_H$ are stabilized subject to the constraint:

$$\begin{bmatrix} \Delta \dot{n}_L \\ \Delta \dot{n}_H \end{bmatrix} = \begin{bmatrix} g_1(\Delta n_L, \Delta n_H) \\ g_2(\Delta n_L, \Delta n_H) \end{bmatrix} + \begin{bmatrix} h(\Delta n_L, \Delta n_H) \\ 0 \end{bmatrix} \tag{3.84}$$

together with the non-zero initial condition $\Delta n_{L0}$ and $\Delta n_{H0}$.

**Remark 4** This is a very challenging control design problem due to the under-actuation of control action. To the best of our knowledge, this problem has not been solved within the finite time control literature, and in fact, lack of the number of actuators or sensors represents an important class of problems to be addressed for any control design methodology.

To solve the above problem, the following conditions are assumed:

***Assumption 1***: a positive constant $\lambda_1$ can be found such that:

$$\|g_2(\Delta n_L, \Delta n_H) - g_2(0, \Delta n_H)\| \leq \lambda_1 \|\Delta n_L\| \tag{3.85}$$

***Assumption 2***: there exist a Lyapunov function $V_0(\Delta n_H)$ and two positive numbers $\lambda_2$, $\lambda_3$ such that:

$$\dot{V}_0(\Delta n_H) = \frac{\partial V_0(\Delta n_H)}{\partial n_H} g_2(0, \Delta n_H) \leq -\lambda_2 \|\Delta n_H\|^2 \tag{3.86}$$

$$\left\| \frac{\partial V_0(\Delta n_H)}{\partial n_H} \right\| \leq \lambda_3 \|\Delta n_H\| \tag{3.87}$$

***Assumption 3***: a positive constant $\lambda_4$ can be found such that:

$$\|g_1(\Delta n_L, \Delta n_H)\| \leq \lambda_4 \|\Delta n_L\| \tag{3.88}$$

**Remark 5** In the above assumptions, assumption 1 implies that the function $g_2(*, \Delta n_H)$ is Lipschizian while assumption 2 says that high-pressure spool is aerodynamically stable; assumption 3 denotes the smoothness and boundedness of low-pressure spool response. These assumptions are valid for turbofan engines due to the fact that the engine is essentially an inertial system at any operating condition.

Now with the above assumptions, the problem can be addressed by the following proposition:

**Proposition 2** *For the turbofan engine model* (3.83),*the static state feedback control*

$$\begin{array}{ll} \Delta W_f = k(\Delta n_L, \Delta n_H) - \frac{cV^{\eta}(\Delta n_L, \Delta n_H)}{\Delta n_L}, & \text{if } \Delta n_L \neq 0 \\ \Delta W_f = 0 & \text{if } \Delta n_L = 0 \end{array} \tag{3.89}$$

*stabilizes the closed-loop system in a finite-time, provided that the following condition holds:*

$$\Delta n_L g_1(\Delta n_L, \Delta n_H) + \Delta n_L k(\Delta n_L, \Delta n_H) + \frac{\partial V_0(\Delta n_H)}{\partial n_H} g_2(\Delta n_L, \Delta n_H) \leq 0 \tag{3.90}$$

*where* $k(\Delta n_L, \Delta n_H)$*is a polynomial in* $\Delta n_L$*and* $\Delta n_H$, *and* $V(\Delta n_L, \Delta n_H) = \frac{1}{2}\Delta n_L^2 + V_0(\Delta n_H)$.

***Proof*** Consider the Lyapunov function candidate $V(\Delta n_L, \Delta n_H) = \frac{1}{2}\Delta n_L^2 + V_0(\Delta n_H)$. Differentiation along the trajectory of (3.83) leads to:

$$\dot{V}(\Delta n_L, \Delta n_H) = \Delta n_L \Delta \dot{n}_L + \frac{\partial V_0(\Delta n_H)}{\partial n_H} g_2(\Delta n_L, \Delta n_H)$$

Substituting the engine dynamics into the above expression results in:

$$\begin{aligned} \dot{V}(\Delta n_L, \Delta n_H) &= \Delta n_L \left[ g_1(\Delta n_L, \Delta n_H) + k(\Delta n_L, \Delta n_H) - \frac{cV^{\eta}(\Delta n_L, \Delta n_H)}{\Delta n_L} \right] \\ &\quad + \frac{\partial V_0(\Delta n_H)}{\partial n_H} g_2(\Delta n_L, \Delta n_H) \\ &= \Delta n_L g_1(\Delta n_L, \Delta n_H) + \Delta n_L k(\Delta n_L, \Delta n_H) - cV^{\eta}(\Delta n_L, \Delta n_H) \\ &\quad + \frac{\partial V_0(\Delta n_H)}{\partial n_H} g_2(\Delta n_L, \Delta n_H) \end{aligned}$$

Use of condition (3.90) in Proposition 2 gives $\dot{V}(\Delta n_L, \Delta n_H) \leq -cV^{\eta}(\Delta n_L, \Delta n_H)$.

Hence from Proposition 1, the closed-loop system is guaranteed to be stabilized in a finite time $T = t_0 + \frac{V^{1-\eta}(t_0)}{c(1-\eta)}$. (END).

The above result is a simple application of finite time stability theory but it provides a framework for control design. In specific, for practical application it is desirable to require the control signal to be generated relatively simply, e.g. $k(\Delta n_L, \Delta n_H)$ is linear in the controlled variable $\Delta n_L$. This is provided by the following important result:

**Proposition 3** *For the turbofan engine model* (3.83),*the static state feedback control:*

$$\begin{aligned} \Delta W_f &= k_1 \Delta n_L - \tfrac{cV^{\eta}(\Delta n_L, \Delta n_H)}{\Delta n_L}, && \text{if } \Delta n_L \neq 0 \\ \Delta W_f &= 0 && \text{if } \Delta n_L = 0 \end{aligned} \tag{3.91}$$

*stabilizes the closed-loop system in a finite-time, provided that the static gain* $k_1$*satisfies the following condition:*

$$k_1 \leq -\lambda_4 - \frac{\lambda_1^2 \lambda_3^2}{4\lambda_2} \tag{3.92}$$

***Proof*** Substitute $k(\Delta n_L, \Delta n_H) = k_1 \Delta n_L$ into condition (3.90):

$$\Delta n_L g_1(\Delta n_L, \Delta n_H) + k_1 \|\Delta n_L\|^2 + \frac{\partial V_0(\Delta n_H)}{\partial n_H} g_2(\Delta n_L, \Delta n_H) \leq 0 \tag{3.93}$$

$$\begin{aligned} &\Delta n_L g_1(\Delta n_L, \Delta n_H) + k_1 \|\Delta n_L\|^2 + \frac{\partial V_0(\Delta n_H)}{\partial n_H} [g_2(\Delta n_L, \Delta n_H - g_2(0, \Delta n_H)] \\ &+ \frac{\partial V_0(\Delta n_H)}{\partial n_H} g_2(0, \Delta n_H) \leq 0 \end{aligned} \tag{3.94}$$

Consider conditions (3.85)–(3.88), after a manipulation of Eqs. (3.93) and (3.94), it is then obtained:

$$\Delta n_L g_1(\Delta n_L, \Delta n_H) \leq \lambda_4 \|\Delta n_L\|^2 \tag{3.95}$$

$$\frac{\partial V_0(\Delta n_H)}{\partial n_H} [g_2(\Delta n_L, \Delta n_H - g_2(0, \Delta n_H)] \leq \lambda_1 \lambda_3 \|\Delta n_L\| \|\Delta n_H\| \tag{3.96}$$

Thus inequality (3.94) will be satisfied if the following condition holds:

$$\lambda_4 \|\Delta n_L\|^2 + k_1 \|\Delta n_L\|^2 + \lambda_1 \lambda_3 \|\Delta n_L\| \|\Delta n_H\| - \lambda_2 \|\Delta n_H\|^2 \leq 0 \tag{3.97}$$

An application of the technique of completing squares leads to the conclusion that the above condition will hold if $\lambda_1^2 \lambda_3^2 + 4\lambda_2(k_1 + \lambda_4) \leq 0$, thus is the condition (3.92).

Therefore, if the static gain $k_1$ is chosen that satisfies condition (3.92), then the Lyapunov function candidate $V(z_1, z_2) = \frac{1}{2} z_1^2 + V_0(z_2)$ along the engine dynamics will have $\dot{V}(t) \leq -cV^{\eta}(t)$. Thus, finite time stabilization is achieved via the static state feedback control (3.91). (END).

Now it is noted that the finite time control law in Eq. (3.91) contains the measurement of both high-pressure spool speed $\Delta n_H$ and low-pressure spool speed $\Delta n_L$. As nowadays engine control systems are increasingly integrated with EHMS (engine health management systems), one of the important methods for EHMS is to use the health condition of one spool speed to estimate the health condition of the other. Hence it will not be desirable to use two measurements simultaneously for control purpose.[1] To address this problem, it is noted that a turbofan engine is an inertial system that is itself a stable dynamical system. Thus, one proposal can be put forward to control only one spool speed with only one spool speed measurement as well. This is the static gain, single sensor, and single actuator approach to turbofan engine control design. The result is summarized in the following important proposition.

**Proposition 4** *For the turbofan engine model* (3.83),*the one spool speed control (single sensor measurement and single actuator control):*

$$\begin{aligned} \Delta W_f &= k_1 \Delta n_L - \frac{c\overline{V}^{\eta}(\Delta n_L)}{\Delta n_L}, && \text{if } \Delta n_L \neq 0 \\ \Delta W_f &= 0 && \text{if } \Delta n_L = 0 \end{aligned} \tag{3.98}$$

*stabilizes the closed-loop system in a finite time, provided that the static gain* $k_1$*satisfies the following condition:*

$$k_1 \leq -\lambda_4 - \frac{\lambda_1^2 \lambda_3^2}{4\lambda_2} \tag{3.92}$$

***Proof*** Now it is essentially considering the problem of finite time regulation of the dynamical system: $\Delta \dot{n}_L = g_1(\Delta n_L, \Delta n_H) + \Delta W_f$. By considering the Lyapunov function candidate $\overline{V}(\Delta n_L) = \frac{1}{2}\Delta n_L^2$ and substituting $k(\Delta n_L, \Delta n_H) = k_1 \Delta n_L$, re-iteration of the process of the above proofs leads to the conclusion that the closed-loop system will be finite timely regulated if $(\lambda_4 + k_1)\|\Delta n_L\|^2 \leq 0$. That is: $k_1 \leq -\lambda_4$. Due to the non-negativeness of constants $\lambda_1 - \lambda_4$, it is seen that satisfaction of condition (3.92) will provide a finite time regulation of the closed-loop system. Noting that condition (3.92) introduces further conservativeness than the condition $k_1 \leq -\lambda_4$, this is for demonstration of the fact that the same parameter can be chosen for one spool control. (END).

**Remark 6** Comparing with controller (3.91), it is seen that for the same static gain $k_1$, controller (3.98) utilizes only the measurement of $\Delta n_L$. This one spool speed control approach to turbofan engine control design solves the challenging problem of non-simultaneous controllability of $\Delta n_H$ and $\Delta n_L$, while the measurement of

[1] It is noted that the concept of using health condition of $\Delta n_H$ (or $\Delta n_L$) to estimate the health condition of $\Delta n_L$ (or $\Delta n_H$) has been an important issue associated with engine fault monitoring and detection, particularly for enhancing the reliability and fault tolerance capability of EHMS.

$\Delta n_H$ can be further used to enhance the capability of EHMS representing one of the important research directions for integration of control and health management systems.

### 3.5.3 Disturbance Attenuation and Robustness

One of the important problems for regulation is disturbance attenuation, e.g. inlet distortion due to environmental change causes thermo-mechanical cycle drift along gas path of the engine, and the set point controller has to counteract this disturbance on shaft rotational speed to maintain the thrust. To investigate the robustness property, now consider the engine model (3.83) with uncertainty and disturbance to both channels:

$$\begin{bmatrix} \Delta\dot{n}_L \\ \Delta\dot{n}_H \end{bmatrix} = \begin{bmatrix} g_1(\Delta n_L, \Delta n_H) \\ g_2(\Delta n_L, \Delta n_H) \end{bmatrix} + \begin{bmatrix} \Delta g_1(\Delta n_L, \Delta n_H) \\ \Delta g_2(\Delta n_L, \Delta n_H) \end{bmatrix} + \begin{bmatrix} \Delta W_f \\ 0 \end{bmatrix} + \begin{bmatrix} d'_L(t) \\ d'_H(t) \end{bmatrix} \tag{3.99}$$

where $\Delta g_1(\Delta n_L, \Delta n_H)$ and $\Delta g_2(\Delta n_L, \Delta n_H)$ are uncertainties associated with the dynamics of $g_1(\Delta n_L, \Delta n_H)$ and $g_2(\Delta n_L, \Delta n_H)$, respectively; $d'_L(t)$ and $d'_H(t)$ are the corresponding disturbances. To consider the disturbance attenuation and robustness properties, $\Delta g_1(\Delta n_L, \Delta n_H)$, $\Delta g_2(\Delta n_L, \Delta n_H)$, $d'_L(t)$ and $d'_H(t)$ are usually assumed to be bounded functions. Therefore, they can be "lumped" together to simplify the analysis while still retaining the key issue: to investigate the effect of bounded disturbance or bounded uncertainty on control system performance. Hence rewrite the above model as:

$$\begin{bmatrix} \Delta\dot{n}_L \\ \Delta\dot{n}_H \end{bmatrix} = \begin{bmatrix} g_1(\Delta n_L, \Delta n_H) \\ g_2(\Delta n_L, \Delta n_H) \end{bmatrix} + \begin{bmatrix} \Delta W_f \\ 0 \end{bmatrix} + \begin{bmatrix} d_L(t) \\ d_H(t) \end{bmatrix} \tag{3.100}$$

where $d_L(t)$ and $d_H(t)$ represent the disturbance and/or uncertainty to each channel with:

$$\|d_L(t)\| \le l_1 \text{ and } \|d_H(t)\| \le l_2 \tag{3.101}$$

The following important result can be proved:

**Proposition 5** *For the turbofan engine model* (3.100)*with uncertainty and disturbance* (3.101),*the one spool speed control (single sensor measurement and single actuator control):*

$$\Delta W_f = k_1 \Delta n_L - \frac{c\overline{V}^{\eta}(\Delta n_L)}{\Delta n_L},\ with\ k_1 \le -\lambda_4 - \frac{\lambda_1^2\lambda_3^2}{4\lambda_2},\ if\ \Delta n_L \ne 0 \\ \Delta W_f = 0\ if\ \Delta n_L = 0 \tag{3.102}$$

*stabilizes the low-pressure shaft speed* $\Delta n_L$ *in a finite-time to a region* $Q$ *defined by* $Q = \left\{\Delta n_L : |\Delta n_L| \le \frac{4\lambda_2 l_1}{\lambda_1^2\lambda_3^2}\right\}$, *and the high-pressure shaft speed* $\Delta n_H$ *will be able to tolerate a bounded uncertainty or disturbance up to* $l_2 \le \frac{\lambda_2}{\lambda_3}\|\Delta n_H\|$.

***Proof*** Consider the low-pressure and high-pressure shaft dynamics separately. For low-pressure shaft, choose the Lyapunov function candidate $\overline{V}(\Delta n_L) = \frac{1}{2}\Delta n_L^2$ and differentiate along the dynamics $\Delta\dot{n}_L = g_1(\Delta n_L, \Delta n_H) + \Delta W_f + d_L(t)$.

$$\begin{aligned}\dot{\overline{V}}(\Delta n_L) &= \Delta n_L\left[g_1(\Delta n_L, \Delta n_H) + k_1\Delta n_L - \frac{c\overline{V}^{\eta}(\Delta n_L)}{\Delta n_L} + d_L(t)\right] \\ &\le (k_1 + \lambda_4)\|\Delta n_L\|^2 + l_1\|\Delta n_L\| - c\overline{V}^{\eta}(\Delta n_L)\end{aligned}$$

Let $Q = \left\{\Delta n_L : |\Delta n_L| \le \frac{4\lambda_2 l_1}{\lambda_1^2\lambda_3^2}\right\}$, for any $\Delta n_L \in \Re - Q$, there has $|\Delta n_L| > \frac{4\lambda_2 l_1}{\lambda_1^2\lambda_3^2}$. That is:

$(k_1 + \lambda_4)\|\Delta n_L\|^2 < -l_1\|\Delta n_L\|$, or $(k_1 + \lambda_4)\|\Delta n_L\|^2 + l_1\|\Delta n_L\| < 0$. Thus: $\dot{\overline{V}}(\Delta n_L) \le (k_1 + \lambda_4)\|\Delta n_L\|^2 + l_1\|\Delta n_L\| - c\overline{V}^{\eta}(\Delta n_L) < -c\overline{V}^{\eta}(\Delta n_L)$. Hence the low-pressure shaft speed $\Delta n_L$ is stabilized in a finite-time to region $Q$.

Now consider the high-pressure shaft dynamics, it is known:

$$\begin{aligned}\dot{V}_0(\Delta n_H) &= \frac{\partial V_0(\Delta n_H)}{\partial n_H}\Delta\dot{n}_H = \frac{\partial V_0(\Delta n_H)}{\partial n_H}[g_2(\Delta n_L, \Delta n_H - g_2(0, \Delta n_H)] \\ &+ \frac{\partial V_0(\Delta n_H)}{\partial n_H}[g_2(0, \Delta n_H) + d_H(t)] \\ &\le \lambda_1\lambda_3\|\Delta n_L\|\|\Delta n_H\| - \lambda_2\|\Delta n_H\|^2 + l_2\lambda_3\|\Delta n_H\|\end{aligned}$$

Now if $(\lambda_1\|\Delta n_L\| + l_2)\lambda_3 - \lambda_2\|\Delta n_H\| \le 0$, that is: $l_2 \le \frac{\lambda_2}{\lambda_3}\|\Delta n_H\| - \lambda_1\|\Delta n_L\|$, then we will have $\dot{V}_0(\Delta n_H) \le 0$, assuming for the moment that $\frac{\lambda_2}{\lambda_3}\|\Delta n_H\| - \lambda_1\|\Delta n_L\| > 0$ due to the fact that $l_2 > 0$. Now consider the region $Q = \left\{\Delta n_L : |\Delta n_L| \le \frac{4\lambda_2 l_1}{\lambda_1^2\lambda_3^2}\right\}$, as the low-pressure shaft speed $\Delta n_L$ is stabilized in a finite-time to region $Q$, hence within $Q$, $\Delta n_L$ will converge to zero after the finite time $T = t_0 + \frac{V^{1-\eta}(t_0)}{c(1-\eta)}$. That is, after $T$, the condition $l_2 \le \frac{\lambda_2}{\lambda_3}\|\Delta n_H\| - \lambda_1\|\Delta n_L\|$ becomes $l_2 \le \frac{\lambda_2}{\lambda_3}\|\Delta n_H\|$. This condition leads to $\dot{V}_0(\Delta n_H) \le 0$, proving the claim in the Proposition. (END).

**Remark 7** It is shown that the one spool speed control (3.102) (single sensor measurement and single actuator control) finite-timely stabilizes the low-pressure shaft speed into a region $Q$ other than the origin under persistent yet bounded disturbance. The area of the region depends on the uncertainty bound $l_1$ while within the region, the ability of tolerating disturbance for the high-pressure shaft speed is proportional to the relative magnitude of $\Delta n_H$.

### 3.5.4 Finite Time Set Point Control of Aeroengines

Now consider the following turbofan engine model around the cruise condition represented by:

$$\begin{aligned}\Delta \dot{n}_L &= -4.1476\Delta n_L + 1.4108\Delta n_H + 12\Delta n_L^2 - \Delta n_H^2 + \Delta w_f \\ \Delta \dot{n}_2 &= 0.2975\Delta n_L - 3.1244\Delta n_H - 1.7\Delta n_L^2 + \Delta n_H^2\end{aligned} \tag{3.103}$$

The nonlinear model has been taken to investigating the advanced control concepts, comparing with linear controls almost exclusively utilized in the state-of-the-art engine control design.

***Finite Time Set Point Control of Aeroengines***

It is first noted that $\Delta \dot{n}_H = g_2(\Delta n_L, \Delta n_H)$ is smooth in a neighbourhood of $\Delta n_H = 0$ and $\Delta \dot{n}_H = g_2(0, \Delta n_H)$ is uniformly exponentially stable around $\Delta n_H = 0$. The assumptions made in Sect. 3.5.2 are satisfied. Now take $V(\Delta n_L, \Delta n_H) = \frac{1}{2}\Delta n_L^2 + V_0(\Delta n_H) = \frac{1}{2}\left(\Delta n_L^2 + \Delta n_H^2\right)$ and from Proposition 2, the controller (3.89) will stabilize the engine in a finite time if the $k(\Delta n_L, \Delta n_H)$ is such chosen that the following condition is satisfied:

$$\Delta n_L g_1(\Delta n_L, \Delta n_H) + \Delta n_L k(\Delta n_L, \Delta n_H) + \Delta n_H g_2(\Delta n_L, \Delta n_H) \le 0 \tag{3.104}$$

Now with a static gain $k(\Delta n_L, \Delta n_H) = k_1 \Delta n_L$, condition (3.104) leads to:

$$\begin{aligned}&(k_1 - 4.1476 - 1.7\Delta n_H + 12\Delta n_L)\Delta n_L^2 + 1.7083\Delta n_L \Delta n_H \\ &\quad + (\Delta n_H - \Delta n_L - 3.1244) \le 0\end{aligned} \tag{3.105}$$

Considering the bounded incremental change of $|\Delta n_L| \le 1$ and $|\Delta n_H| \le 1$, inequality (3.105) will be satisfied if the gain $k_1$ satisfies:

$$k_1 < -10.2013 \tag{3.106}$$

Take $k_1 = -11$ and the static state feedback control is given by:

$$\begin{aligned}\Delta W_f &= -11\Delta n_L - \tfrac{cV^{\eta}(\Delta n_L, \Delta n_H)}{\Delta n_L}, && \text{if } \Delta n_L \ne 0 \\ \Delta W_f &= 0 && \text{if } \Delta n_L = 0\end{aligned} \tag{3.107}$$

From Proposition 3, it is claimed that the controller (3.107) will regulate the engine in a finite time. Now choose the initial conditions, e.g. $\left[n_L(0)\ n_H(0)\right] = \left[-0.5\ 0.8\right]$, constant $c = 2$, and $\eta = 0.5$. Figures 3.33 and 3.34 show the simulation

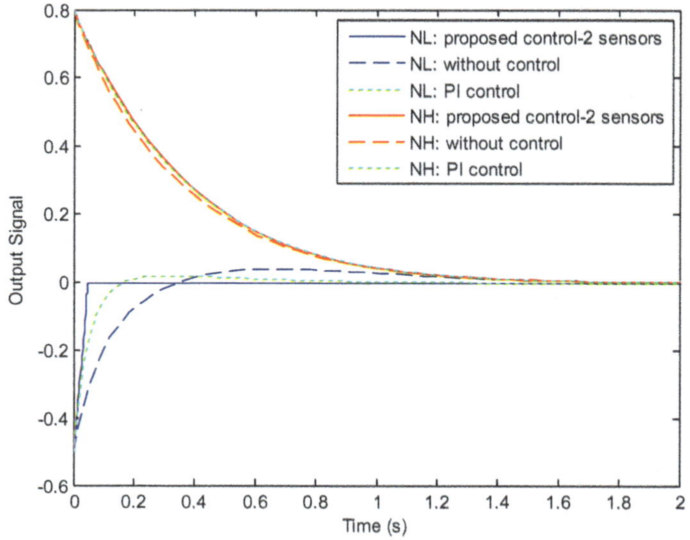


**Fig. 3.33** Performance of the finite-time control with 2 sensor measurements and PI control

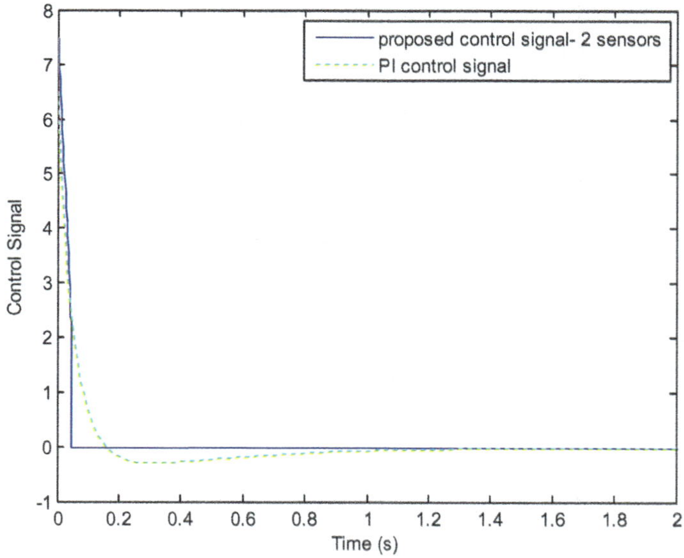


**Fig. 3.34** Control signals for both finite-time control with 2 sensor measurements and PI control

results for the proposed control, also shown is the performance of the corresponding PI control. It can be seen that the system can be stabilized in finite time using the proposed control with a short transient response, clearly validating the above claims.

Now noting that implementation of controller (3.107) will require measurements from both $\Delta n_H$ and $\Delta n_L$. And it has been remarked (see the footnote in Sect. 3.5.2) that, for EHMS purpose, engine control systems utilize the health condition of one spool speed to estimate the health condition of the other to increase system reliability. Fortunately Proposition 4 says that the single sensor measurement and single actuator control can also stabilize the closed-loop system in a finite time:

$$\begin{array}{ll} \Delta W_f = k_1 \Delta n_L - \frac{c\overline{V}^{\eta}(\Delta n_L)}{\Delta n_L}, & \text{if } \Delta n_L \neq 0 \\ \Delta W_f = 0 & \text{if } \Delta n_L = 0 \end{array} \tag{3.108}$$

With the same initial conditions and parameter selection, the single sensor controller performance is shown in Figs. 3.35 and 3.36, also shown is the performance of the 2-sensor measurement controller (3.107). It is seen that controller (3.108) can also stabilize the closed loop systems while the loss of performance is marginal, validating the claim in Proposition 4.

***Robustness of Finite Time Set Point Controller***

The engine consistently subjects to disturbance from both environmental change and fuel metering devices. Thus the proposed controller must be able to tolerate disturbances to provide certain robustness property. Consider the situation where disturbances are injected into both channels with:

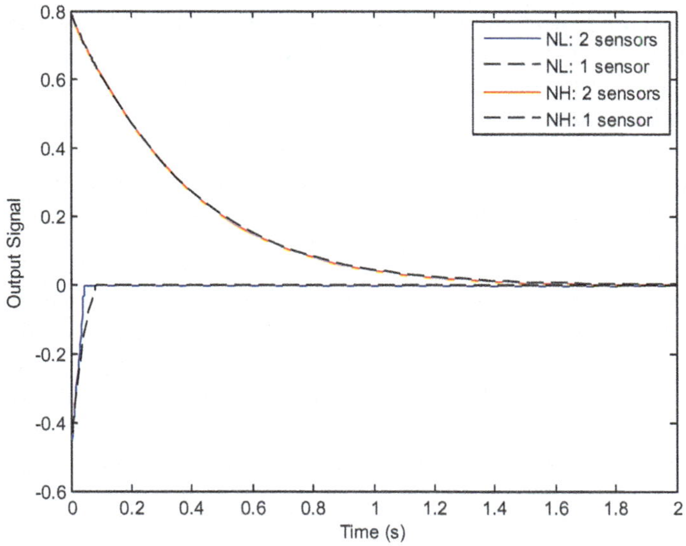


**Fig. 3.35** Performance of the finite-time control: 1 sensor *verses* 2-sensor measurement

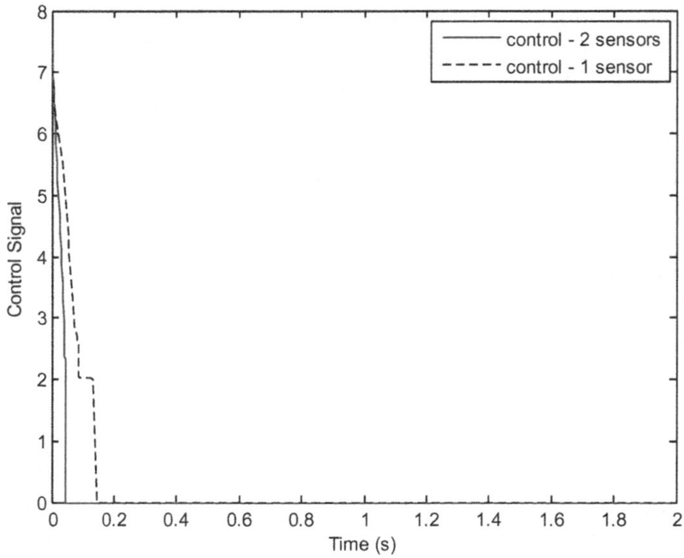


**Fig. 3.36** Control signals for finite-time control: 1 sensor *verses* 2-sensor measurement

$$
\begin{aligned}
d_1(t) &= A\sin(60t) + 0.1U(0,1) \\
d_2(t) &= A\sin(80t) + 0.1U(0,1)
\end{aligned}
\tag{3.109}
$$

The disturbances contain a random noise ($U(0,1)$ denoting a random signal with zero mean and unity variance) superposed by a low frequency signal ($A$ is the magnitude), replicating typical pulsation dynamics of the engine fuel flow regulator and aero-dynamical disturbances. Still consider the single sensor measurement and single actuator controller (3.108), the performance for $A = 2$ is shown in Figs. 3.37 and 3.38. To further experiment with the disturbance attenuation property of the controller (3.108), multiple disturbances with different magnitude have also been shown for comparison in Fig. 3.39. It can be seen that this single sensor measurement and single actuator controller (3.108) provides finite-time stability and possesses remarkable disturbance tolerance capability.

**Remark 8** Since the low-pressure compressor turbine is only aerodynamically coupled with the high-pressure compressor turbine, regulation performance of $\Delta n_L$ is much better than that of $\Delta n_H$ due to the direct control of fuel flow ratio on $\Delta n_L$. Depending on the control plan, $\Delta n_H$ may also be controlled directly but not both can be controlled simultaneously. Therefore, control through only single actuator is required. And it will be ideal if only one sensor is utilized as well.

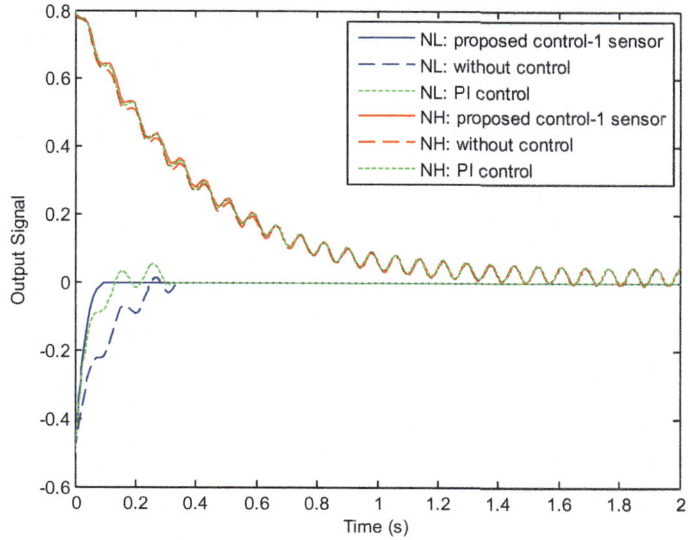


**Fig. 3.37** Performance of the single sensor measurement and single actuator finite-time controller (3.108) under persistent disturbance of magnitude $A = 2$

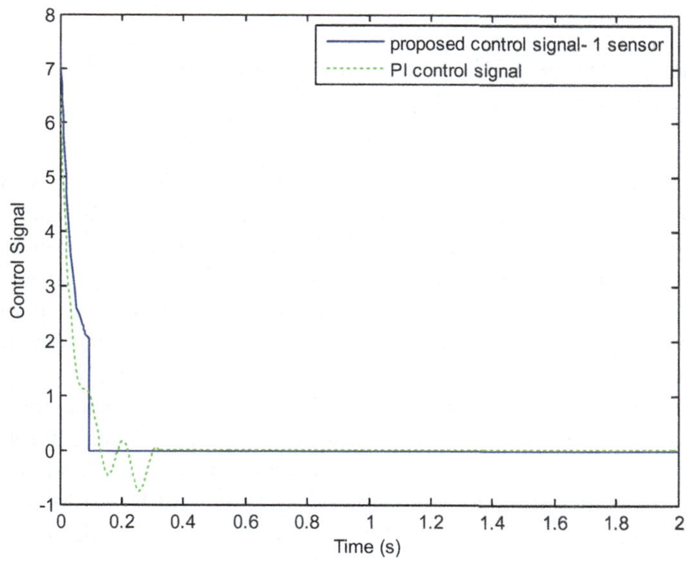


**Fig. 3.38** Control signals for both finite-time controller (3.108) and PI control under disturbances

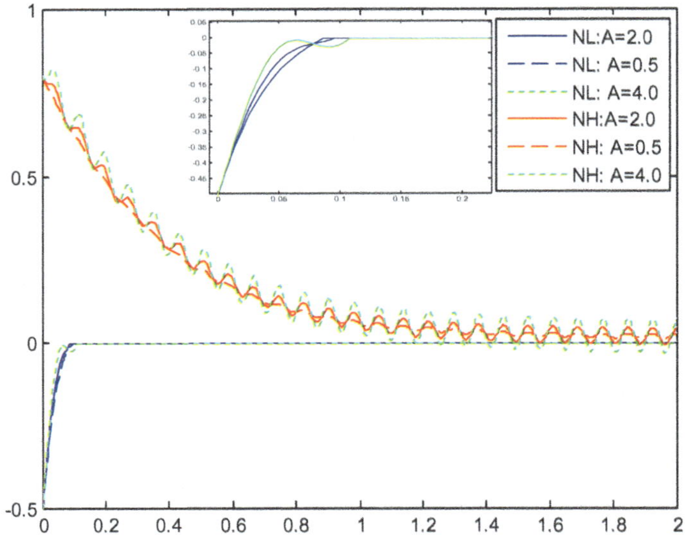


**Fig. 3.39** Performance of the single sensor measurement and single actuator finite-time controller (3.108) under persistent disturbance of different magnitude: left to the legend is the magnified view for low pressure turbine speed $\Delta n_L$

### 3.5.5 Summary and Discussion for Finite Time Set Point Control of Aeroengines

A finite time set point control method has been proposed in this section. A number of important results concerning upon single sensor measurement and single actuator control design problem have been obtained. The problem of disturbance tolerance has also been analyzed. Simulation results have validated the proposed design methodology. Although it can be argued that measurement of both low- and high-pressure turbine rotational speeds is available, control design can indeed be able to proceed through the feedback of both speed values. There does arise the requirement to utilize only one measurement and the other one can be used for health management to infer the faults of the other, henceforth enhancing the reliability of the control systems. Thus the result presented here should contribute to this important field of research.

Meanwhile as the turbofan engine control systems have been researched experiencing "digitalization" process in China over the past three decades, most of the commercial engines are controlled by FADECs (Full Authority Digital Engine Control). For the time being, however, FADECs are still in the development stage in China. There are many facets to be investigated and one of them is the advanced control algorithms that specifically target the nonlinear dynamics of the engine model. The increasing computational capability with high reliability built in the FADEC makes it possible to realize nonlinear control designs with fast response and high performance.

## 3.6 Summary

This chapter has presented a series of methodologies for aeroengine set point control, and although it is conventionally thought of as a simple matter, cares should also be taken, e.g. fast regulation to set point while guaranteeing certain regulation performance, and of course, respecting the corresponding operational limits. This may not be easy as it appears particularly when one is tackling nonlinear set point control problems. The result presented here should contribute to this important field of research.

## References


1. Doyle, J.C. 1984. *Lectures Notes on Advances in Multivariable Control.* Minneapolis: ONR/Honeywell Workshop.
2. Maciejowski, J.M. 1989. *Multivariable Feedback Design.* Addison-Wesley.
3. Doyle, J.C., K. Glover, P.P. Khargonekar, and B.A. Francis. 1989. State-space solutions to standard and control problems. *IEEE Transactions on Automatic Control* 34 (8): 831–847.
4. Glover, K., and J.C. Doyle. 1988. State-space formulae for all stabilising controller that satisfy an norm bound and relations to risk sensitivity. *Systems and Control Letters* 11: 167–172.
5. Hatonen, J. 2005. *Adaptive and Self-Tuning Control: Part I Lecture Notes.* The University of Sheffield.
6. Åström, K.J., and B. Wittenmark. 1995. *Adaptive Control*, 2nd ed. Addison-Wesley.
7. Isidori, A. 1995. *Nonlinear Control Systems*, 3rd ed. Berlin: Springer-Verlag.
8. Van der Shaft, A.J. 1996. *L2 Gain and Passivity Techniques in Nonlinear Control.* Heidelberg: Springer-Verlag.
9. Byrnes C.I. 2000. Toward a nonequilibrium theory for nonlinear control systems. In *Lecture Notes in Control and Information Sciences.* Berlin: Springer.
10. Sepulchre, R., M. Jankovic, and P.V. Kokotovic. 1997. *Constructive Nonlinear Control.* New York: Springer-Verlag.
11. Block, A.M., J. Baillieul, P. Crouch, and J. Marsden. 2003. *Nonholonomic Mechanics and Control.* Springer.
12. Mayne, D.Q., J.B. Rawlings, C.V. Rao, and P.O.M. Scokaet. 2000. Constrained model predictive control: Stability and optimality. *Automatica* 36: 787–814.
13. Zhou, K., J.C. Doyle, and K. Glover. 1996. *Robust and Optimal Control.* Prentice Hall.
14. Kristić, M., J. Sun, and P.V. Kokotović. 1996. Robust control of nonlinear systems with input unmodeled dynamics. *IEEE Transactions on Automatic Control* 41: 913–920.
15. Ahmadi, M., H. Mojallali, and R. Wisniewski. 2012. Robust $H\infty$ control of uncertain switched systems defined on polyhedral sets with Filippov solutions. *ISA Transactions* 51 (6): 722–731.
16. de Best, J., B. Bukkems, M. van de Molengraft, W. Heemels, and M. Steinbuch. 2008. Robust control of piecewise linear systems: A case study in sheet flow control. *Control Engineering Practice* 16: 991–1003.
17. Franklin, G.F., J.D. Powell, and A. Emami-Naeini. 2002. *Feedback Control of Dynamical Systems*, 4th ed. Addison-Wesley Longman Publishing Co. Inc.
18. Radke, A., and Z. Gao. 2006. A survey of state and disturbance observers for practitioners. In *Proceedings of the 2006 American Control Conference*, 5183–5188. 1-42440209-3/06, USA, Minnesota.
19. Li, S., J. Yang, W. Chen, and X. Chen. 2014. *Disturbance Observer-Based Control: Methods and Applications.* CRC Press.

20. Choi, Y., K. Yang, W.K. Chung, et al. 2003. On the robustness and performance of disturbance observers for second-order systems. *IEEE Transactions on Automatic Control* 48 (2): 315–320.
21. Choi, H.T., B.K. Kim, I.H. Suh, et al. 2000. Design of robust high-speed motion controller for a plant with actuator saturation. *Transactions of the ASME Journal of Dynamic Systems, Measurement and Control* 122 (3): 535–541.
22. Lee, K.B., and F. Blaabjerg. 2007. Robust and stable disturbance observer of servo system for low-speed operation. *IEEE Transactions on Industry Applications* 43 (3): 627–635.
23. Yang, J., A. Zolotas, W. Chen, K. Michail, and S. Li. 2011. Robust control of nonlinear MAGLEV suspension system with mismatched uncertainties via DOBC approach. *ISA Transactions* 50 (3): 389–396.
24. Han, J. 2009. From PID to active disturbance rejection control. *IEEE Transactions on Industrial Electronics* 56 (3): 900–906.
25. Yang, Gang, Jianguo Sun, Hua Yao, Jun Zang, Aiping Liu, Xuefei Yi, and Lihui Zhang. 2006. Experimental verification of H∞/LTR method for aeroengine control systems. *Acta Aeronautica et Astronautica Sinica* 27 (5): 773–777 (In Chinese).
26. Härefors, M. 1997. Application of H∞ robust control to the RM12 jet engine. *Control Engineering Practice* 5 (9): 1189–1201.
27. Frederick, D.K., S. Garg, and S. Adibhatla. 2000. Turbofan engine control design using robust multivariable control technologies. *IEEE Transactions on Control Systems Technology* 8 (6): 961–970.
28. Wang, H., Y. Guo, and G, Li. 2008. Aero-engine control using two-degree-of-freedom H∞ approach, 1-4244-2386-6. IEEE.
29. Pachpatte, B. 1973. A note on Gronwall-Bellman inequality. *Journal of Mathematical Analysis and Applications* 44: 758–762.
30. Desoer, C., and M. Viddyasagar. 1975. *Feedback Systems: Input-Output Properties*. NY: Academic Press.
31. Viddyasagar, M. 1993. *Nonlinear Systems Analysis*, 2nd ed. NJ: Prentice Hall.
32. Shimizu, K. 2000. Nonlinear state observers by gradient descent method. In *Proceedings of IEEE Coneference on Control and Applications*. Anchorage, USA.
33. Pachpatte, B. 1975. On some generalizations of Bellmans'lemma. *Journal of Mathematical Analysis and Applications* 51: 141–150.
34. Pachpatte, B. 1975. On some integral inequalities similar to Bellman-Bihari inequalities. *Journal of Mathematical Analysis and Applications* 49: 794–802.
35. El Alami, N. 1995. A generalization of Gronwall's lemma. In *Proceedings of the International Conference on Differential Equations*. Marrakesh, Morocco.
36. N'Doye, I., M. Zasadzinski, M. Darouach, N. Radhy, and A. Bouaziz. 2011. Exponential stabilization of a class of nonlinear systems: A generalized Gronwall-Bellman lemma approach. *Nonlinear Analysis* 74: 7333–7341.
37. Adibhatla, S., H, Brown, and Z. Gastineau. 1992. Intelligent Engine Control (IEC). In *ALASA 92–384, 28th Joint Propulsion Conference Exhibit*. Nashville, TN.
38. Slotine, J.J., and W. Li. 1991. *Applied Nonlinear Control*. New Jersey: Prentice Hall International Inc.
39. Kokotovic, P., and M. Arcak. 2001. Constructive nonlinear control: A historical perspective. *Automatica* 37: 637–662.
40. Wang, Y., and Y. Guo. 2006. Optimal PID control system design on aero-engine using LQR approach. *Journal of Changchun University of Science and Technology* 29 (1): 122–124.
41. Shang, G. 2014. *Aircraft Engine Control Based on Nonlinear Models*. MSc. diss., Nanjing University of Aeronautics and Astronautics.
42. Syrmos, V.L., C.T. Abdallah, P. Dorato, and K. Grigoriadis. 1997. Static output feedback: A survey. *Automatica* 33: 125–137.
43. Overton, M.L., and R.S. Womersley. 1988. On minimizing the spectral radius of a nonsymmetric matrix function—optimality conditions and duality theory. *SIAM Journal on Matrix Analysis and Applications* 9: 474–498.

44. Toscano, R. 2006. A simple method to find a robust output feedback controller by random search approach. *ISA Transactions* 45: 35–44.
45. Kim, Y. 2012. Static output feedback stabilization of interconnected systems. *Systems and Control Letters* 61: 381–386.
46. Kim, Y., D.W. Gu, and I. Postlethwaite. 2009. Spectral radius minimization for optimal average consensus and output feedback stabilization. *Automatica* 45: 1379–1386.
47. Burke, J., A. Lewis, and M. Overton. 2002. Two numerical methods for optimizing matrix stability. *Linear Algebra and its Applications* 351–352: 117–145.
48. Gumussoy, S., D. Henrion, M. Millstone, and M.L. Overton. 2009. Multi-objective robust control with HIFOO 2.0. In *Proceedings of IFAC Symposium on Robust Control Design*. Haifa, Israel.
49. Nguang, S.K., S. Saat, and M. Krug. 2011. Static output feedback controller design for uncertain polynomial systems: An iterative sums of squares approach. *IET Control Theory and Applications* 5: 1079–1084.
50. Parrilo, P.A. 2000. *Structured Semidefinite Programes and Semialgebraic Geometry Methods in Robustness and Optimization*. PhD diss., California Institute of Technology, Pasadena.
51. Jiang, S., and H. Fang. 2013. H$\infty$ static output feedback control for nonlinear networked control systems with time delays and packet dropouts. *ISA Transactions* 52: 215–222.
52. Xia, Z., J. Li, and J. Li. 2012. Delay-dependent fuzzy static output feedback control for discrete-time fuzzy stochastic systems with distributed time-varying delays. *ISA Transactions* 51: 702–712.
53. Astolfi, A., and P. Colaneri. 2002. A Hamilton-Jacobi setup for the static output feedback stabilization of nonlinear systems. *IEEE Transactions on Automatic Control* 47: 2038–2041.
54. Astolfi, A., and P. Colaneri. 2000. Static output feedback stabilization of linear and nonlinear systems. In *Proceedings of the 39th IEEE Conference on Decision and Control*, 2920–2925. Sydney, Australia.
55. Sakamoto, N., and A.J. van der Schaft. 2008. Analytical approximation methods for the stabilizing solution of the Hamilton-Jacobi equation. *IEEE Transactions on Automatic Control* 53: 2335–2350.
56. Zhao, D., and J.L. Wang. 2010. Robust static output feedback design for polynomial nonlinear systems. *International Journal of Robust and Nonlinear Control* 20: 1637–1654.
57. Fantoni, I., and R. Lozano. 2001. *Non-linear Control for Underactuated Mechanical Systems*. Springer.
58. Wang, J. 2009. *Active Control of Remote Vibration: a Geometric Approach*. PhD diss., The University of Sheffield, UK.
59. Adibhatla, S., H. Brown, and Z. Gastineau. 1992. Intelligent engine control (IEC). In *ALASA92–384, 28th joint propulsion conference and exhibit*. Nashville, TN.
60. Bhat, S.P., and D.S. Bernstein. 1998. Continuous finite-time stabilization of the translational and rotational double integrators. *IEEE Transactions on Automatic Control* 43 (5): 678–682.
61. Ding, S., and S. Li. 2011. A survey for finite-time control problems. *Control and Decision* 26 (2): 1–10.
62. Dorato, P. 2006. An overview of finite-time stability. In *Current Trends in Nonlinear Systems and Control*, 185–194.
63. Hong, Y., J. Huang, and Y. Xu. 2001. On an output feedback finite-time stabilization problem. *IEEE Transactions on Automatic Control* 46 (2): 305–309.
64. Hong, Y. 2002. Finite-time stabilization and stabilizability of a class of controllable systems. *Systems and Control Letters* 46 (4): 231–236.
65. Seo, S., H. Shim, and J.H. Seo. 2008. Global finite-time stabilization of a nonlinear system using dynamic exponent scaling. In *Proceeding 47th IEEE Conference. Decision and Control*. Cancun, Mexico.
66. Li, S., and Y. Tian. 2007. Finite time stability of cascaded time-varying systems. *International Journal of Control* 80 (4): 646–657.
67. Athans, M., and P.L. Falb. 1985. *Optimal Control: An Introduction to the Theory and Its Applications*. New YorK: McGraw-Hill.

68. Ryan, E.P. 1979. Singular optimal controls for second-order saturating systems. *International Journal of Control* 30 (4): 549–564.
69. Ryan, E.P. 1991. Finite-time stabilization of uncertain nonlinear planar systems. *Dynamics and Control* 1 (1): 83–89.
70. Feng, Y., X.H. Yu, and Z. Man. 2002. Non-singular terminal sliding mode control of rigid manipulators. *Automatica* 38 (9): 2159–2167.
71. Yu, S., and X. Yu. 2000. Robust global terminal sliding mode control of SISO nonlinear uncertain systems. In *IEEE Conference on Decision and Control*, 2198–2203.

# Chapter 4
# Model-Based Aeroengine Nonlinear Transient Control

Transient control refers to the engine in transient operation when it experiences acceleration or deceleration. The requirement for transient control is therefore to have a safe and fast transient response when manoeuvring. Therefore, the optimal operating curve, if plotted on a compressor verusus corrected mass flow rate map, would be the line following closely the surge line for acceleration and burner blow-out limit for deceleration. In practice, however, enough surge and stability margin should be retained and more importantly, it is extremely challenging to determine where the real surge line locates. This is a very important research direction that has not been given enough attention. This is the reason that transient control is much more challenging to design and accounts for approximately 75% of the total control design and development effort [1]!

Referring to Fig. 4.1, transient control is completed with set point control and transient schedules. That is, acceleration/deceleration process starts from one set point to another set point, and combines with an acceleration/deceleration schedule in between. Therefore, issues that should be addressed for transient control include:

(1) How to change the engine's operating condition from one state to another and
(2) How to keep the engine from exceeding operating limits while making these state changes (as represented by power level)
(3) How to provide acceptable transient performance, e.g. rising time, settling time, overshoot etc.

To solve these problems, two approaches are conventionally used: ***schedule-based transient control*** and ***acceleration-based transient control***. Schedule-based transient control takes the following form as shown in Fig. 4.2.

It is seen that control authority starts from one set point controller, handles over to the transient controller through a ***Low-Win logic*** for acceleration and a ***High-Win logic*** for deceleration, before transferring the authority to another set point controller; and the time it takes for the set point controller to hand over the control authority depends on the difference between the beginning and the ending speeds defined by the corresponding set points. In case the difference is small, the schedule may

 
J. Wang et al., *Model-based Nonlinear Control of Aeroengines*,
https://doi.org/10.1007/978-981-16-4453-5_4

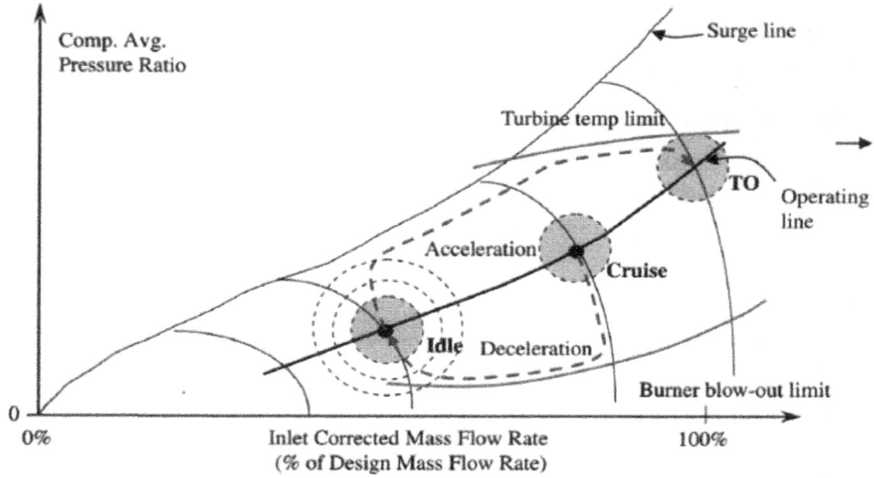


**Fig. 4.1** Transient control as represented in a compressor versus corrected mass flow rate map

**Fig. 4.2** Schedule-based transient control logic

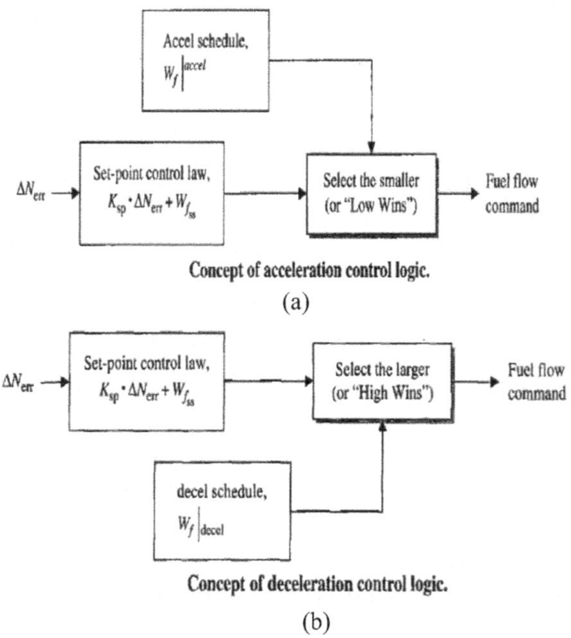


not be activated. To prevent the engine works exceeding its limits, the acceleration and deceleration schedules are carefully tuned and this often results in a lookup table embedded into the engine control system logic. An example of the acceleration schedule takes the following form as in Table 4.1:

**Table 4.1** Engine acceleration schedule and corresponding steady state fuel flow

| $n$ (r/min) | Accel schedule (lb/h) | Steady state fuel flow(lb/h) |
|---|---|---|
| 10,000 | 104 | 100 |
| 25,000 | 116 | 115 |
| 30,000 | 124.4 | 120 |
| 35,000 | 155 | 150 |
| 40,000 | 205 | 200 |
| 45,000 | 329 | 320 |
| 50,000 | 459 | 450 |
| 55,000 | 559 | 550 |

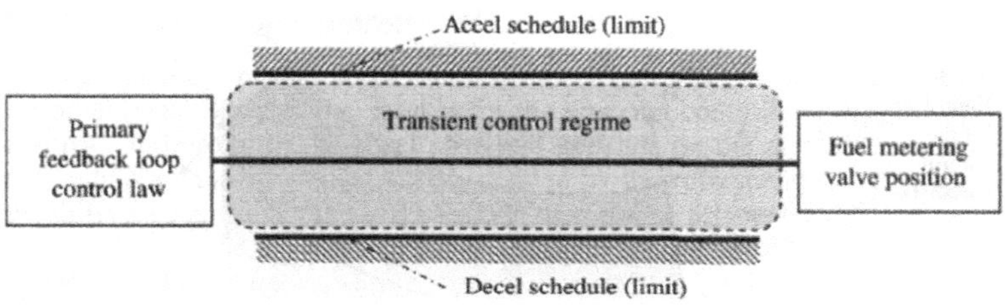


**Fig. 4.3** The schedule should respect all the limits resulting in a control envelope

The above table is obtained through ***extensive simulation and experimental investigation***. The schedule should respect all the operational limits and a schematic illustration is shown in Fig. 4.3.

It is seen that one of the important issues in schedule-based transient control is to choose appropriate scheduling variables. For engine design, the scenario is that the engine operating line is divided into multiple, yet finite number of nominal set points, and then at each nominal point, a set-point controller is designed, which is only valid in the vicinity of the nominal point. The result is that we have a set of linear set point controllers, and a methodology is needed to switch between two neighbouring set point controllers, ***without introducing discontinuities in engine output variables***—gain scheduling! Hence, gain scheduling controller can then be designed through designating one or more state or output variables as the scheduling variables. Then, neighbouring set point controllers are switched according to the value(s) of the scheduling variable(s).

The above schedule-based control solves the transient control design problem but it may lead to sluggish response and for many situations, the engine is required to have a quick response to power level request. In this case, the acceleration-based transient control can be utilized where the engine produces a desired acceleration or deceleration in engine speed directly, rather than modulate the fuel flow to produce a desired shaft speed! The transient logic is defined as:

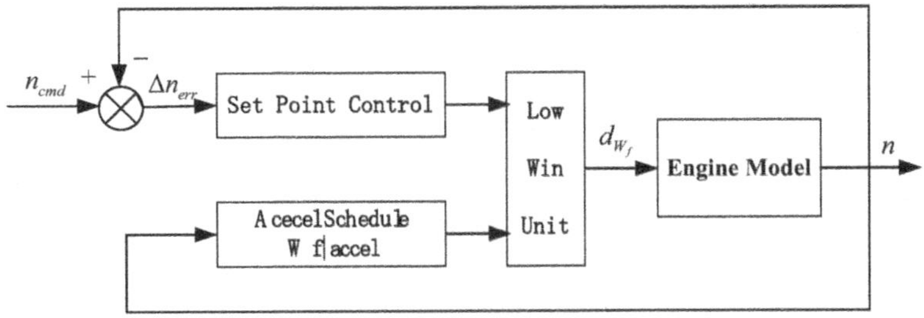


**Fig. 4.4** Acceleration-based transient control

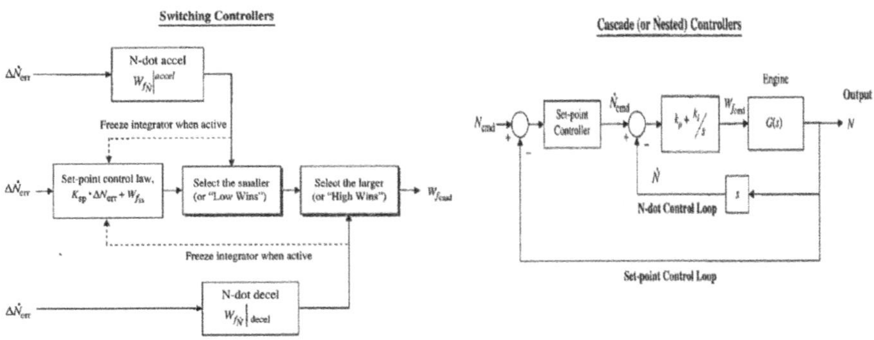


**Fig. 4.5** Two commonly used implementation schemes for acceleration-based control

$$\text{If } K_{sp}\Delta n_{err} < \left( W_f\big|^{accel} - W_{fss} \right), \text{ then } d_{W_f} = K_{sp}\Delta n_{err}$$

$$\text{If } K_{sp}\Delta n_{err} \geq \left( W_f\big|^{accel} - W_{fss} \right), \text{ then } d_{W_f} = W_f\big|^{accel} - W_{fss}$$

where: $d_{W_f}$ is the fuel flow ratio with respect to its steady state value; $W_f\big|^{accel}$ is the fuel flow for acceleration schedule; $W_{fss}$ is the steady state fuel flow value; and $\Delta n_{err} = n_{cmd} - n$, the deviation between the command and measurement for rotational speed of the shaft, and finally $K_{sp}$ is the PI control law for the set point controller.

This acceleration-based control is commonly referred to as the N-dot Control as shown in Fig. 4.4. Two commonly used implementation schemes are also provided in Fig. 4.5.

A comparison of the two transient control approaches is in order [1]:

- ***Advantages of N-dot control:***

(1) consistent transient performance for all of the engines in the fleet regardless of their manufacturing variations, ages and deteriorations;
(2) direct control of the rotor acceleration produces more responsive engine transients;
(3) good stability property for the control loop.

- **Disadvantages:**

(1) tendency to drive the engine into stall, surge or flame-out;
(2) acceleration or deceleration time is not minimized!

For the above transient control, it is seen that the key to transient control design is the transient schedule (acceleration and deceleration schedule) that has been conventionally implemented as lookup tables as shown in Table 4.1. While these gain-scheduling methods have been standard practice in industry, advanced methods are necessitated to further dig up the potentialities of engine performance. However, most of the advanced controls require a mathematical model to carry out control design. In fact, it is well-known that the aero-thermal component-level models can capture the engine dynamics very accurately [2–6]. This type of models can take heat soakage, time delay, turbine cooling etc. into account, making them extremely accurate even within 1% full scale for pressures, temperatures and speeds [7]. However the component-level models are not control-oriented due to the fact the key components such as fan, compressor, and turbine are represented by lookup tables or some other non-analytical modules such as (C or Fortran) program codes, while most of the control theories assume that an analytical model exists before control design. It is illuminating to point out that nonlinear systems theory has developed rapidly over recent decades including concepts such as zero dynamics and normal forms [8], passivity and dissipativity [9], nonequilibrium theory [10] etc. As a consequence, a number of nonlinear control design techniques have been well established such as feedback linearization [8], recursive designs including backstepping and forwarding [11], energy-based control design for nonholonomic dynamical systems [12] and nonlinear model predictive control [13], to name just a few. However, most of the above theories assume implicitly that the system model can be represented by difference/differential equations. Therefore, they can be used for aeroengine set point control, but as mentioned above, for the aeroengines with non-analytical modules, the above methods cannot be applied directly.

To carry out model-based control design, two fundamental approaches can be used to "extract" linear or nonlinear models before the control design process. The first one is linearization, however, due to the local nature of linearization techniques, the model such obtained is only valid within a vicinity of certain operating point (usually within 3–5% of engine shaft speed), and this explains the reason that gain scheduling has been universally adopted in aeroengine control system design, with the most recent development called linear parameter varying control [14–16]. While gain scheduling control has long been a standard practice [17] (particularly for civil aircraft engine controls), its performance can deteriorate seriously during controller switching [18], and this leads to the development of linear parameter varying control to obtain "smooth" transition of controller switching. This approach has been extensively investigated in SNECMA [19–24] and the following challenges have been identified: (1) the linear parameter varying model of aircraft engine is difficult to obtain; (2) control design is relatively difficult to perform due to the non-convexity of LMI; (3) the order of resulting controllers is relatively high. Therefore, linear parameter varying modelling and control needs further investigation particularly for the benefit of on-board real-time control.

The above difficulties associated with the linearization technique to control-oriented modelling are reflected by the recent trend in utilizing system identification approaches. One of the most often used is the neural networks model that constitutes off-line training and on-line adaptation process to improve accuracy and real-time performance, resulting neural networks models over the flight envelope [25–27]. However, to the best of the authors' knowledge, there have not been flight-tested results so far. The significance of neural networks, is that it motivates the utilization of other identification methods for on-board real-time modelling and the investigation of control-oriented MBDs.

The above discussion raises the issue ***whether the engine as a non-analytical system can be controlled directly without linearization or identification of analytical models***. This necessitates the research on control of general nonlinear complex systems. This chapter presents the results on this interesting problem. In particular, Sect. 4.1 considers control design for nonanalytical system representation of aeroengines—direct control design of component-level models. This is motivated by the observation that in aeroengine modelling, the dynamics of compressors, turbines etc. cannot be expressed by a mathematically analytical model such as difference/differential equations; instead, ***they are represented by lookup tables or C/Fortran codes***. Such a treatment regarding the engine as a "black box" is interesting, and its multivariable case for NARX type of model representation is presented in Sect. 4.2; however, Sect. 4.3 shall introduce another switching-based method for performance improvement.

## 4.1 Nonlinear Generalized Minimum Variance Based Aeroengine Transient Control

To solve the aeroengine transient control problem, a block diagram as shown in Fig. 4.6 is considered where the engine can be represented by lookup tables, programming codes or any other non-analytical modules with channel delay. This is a very general representation that replicates the key issues in engine transient control.

To proceed, for the model above, it can be represented using a nonlinear operator:

$$u_0 = W_{1k}(u) \tag{4.1}$$

The sensor model $W_{0k}$ is a linear model (usually a first order inertial element). Together with the time delay, the sensor model can be combined with the disturbance model as:

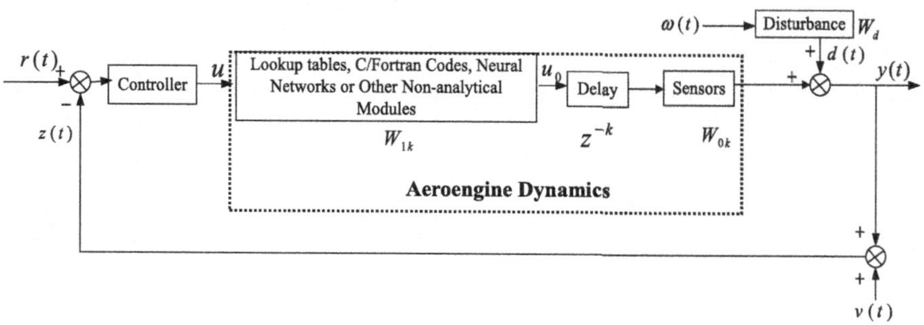


**Fig. 4.6** Block diagram for control of general nonlinear systems

$$
\begin{aligned}
x(t+1) &= Ax(t) + Bu_0(t-k) + E\xi(t) \\
y(t) &= Cx(t) + Du_0(t-k)
\end{aligned}
\tag{4.2}
$$

## 4.1.1 Optimal Controller Design

To derive the optimal control signal, a cost function should first be defined. Here the following minimum variance type of index is considered:

$$J = E\{\phi_0^T(t)\phi_0(t)\} = E\{\mathrm{trace}[\phi_0^T(t)\phi_0(t)]\} \tag{4.3}$$

where: $\phi_0(t) = P_c e(t) + (F_c u)(t)$, and $P_c$ and $F_c$ are design parameters. Considering the signal delay, the control signal can only affect $\phi_0(t)$ after at least $k$ steps. Therefore, there is no loss of generality to express $F_c$ as $F_c = z^{-k}F_{ck}$, then $\phi_0(t)$ can be rewritten as:

$$\phi_0(t+k) = P_c[r(t+k) - y(t+k) - v(t+k)] + (F_{ck}u)(t) \tag{4.4}$$

Now consider the optimization of the cost function $J$, the expression $J = E\{\phi_0^T(t)\phi_0(t)\}$ can be represented by a combination of optimal estimation $\hat{\phi}_0(t+k|t)$ and estimation error $\tilde{\phi}_0(t+k|t)$. An application of orthogonality leads to:

$$J = E\{\phi_0^T(t)\phi_0(t)\} = E\left\{\hat{\phi}_0^T(t+k|t)\hat{\phi}_0(t+k|t)\right\} + E\left\{\hat{\phi}_0^T(t+k|t)\tilde{\phi}_0(t+k|t)\right\} \tag{4.5}$$

The estimation error $\tilde{\phi}_0(t+k|t)$ is not correlated with control input, henceforth the condition for minimizing the cost function $J$ is that the $k$-step estimation $\tilde{\phi}_0(t+k|t) = 0$.

Now:

$$\hat{\phi}_0(t+k|t) = P_c\big[\hat{r}(t+k|t) - C\hat{x}(t+k|t) - Du_0(t)\big] + (F_{ck}u)(t) \tag{4.6}$$

In the above equation, for each component:

$\hat{r}(t+k|t)$: this term is the $k$-step optimal estimation for reference signal. A reference signal can be represented by a linear system with disturbance:

$$\begin{aligned} x_r(t+1) &= A_r x_r(t) + B_r \omega(t) \\ r(t) &= C_r x_r(t) \end{aligned} \tag{4.7}$$

where: $\omega(t)$ is a white noise with zero mean and unity covariance. Therefore, the reference signal $k$-step optimal estimation is:

$$\hat{r}(t+k|t) = C_r^k x_r(t|t) \tag{4.8}$$

$\hat{x}(t+k|t)$: is the $k$-step optimal estimation for the combined model, and thus can be written into:

$$\hat{x}(t+k|t) = A^k \hat{x}(t|t) + T_0(k, z^{-1}) B u_0(t) \tag{4.9}$$

where: $T_0(k, z^{-1}) = z^{-1}\big(I + z^{-1}A + z^{-2}A^2 + \cdots + z^{-(k-1)}A^{k-1}\big)$.

Now it is required to estimate the state $\hat{x}(t|t)$. From the well-known Kalman filter theory, the following state estimation equations are obtained:

$$Predictor\ {:}\hat{x}(t+1|t) = A\hat{x}(t|t) + Bu_0(t-k) \tag{4.10}$$

$$Corrector : \hat{x}(t+1|t+1) = \hat{x}(t+1|t) + K_f\big[e(t+1) - \hat{e}(t+1|t)\big] \tag{4.11}$$

$$\hat{e}(t+1|t) = \hat{r}(t+1|t) - \hat{y}(t+1|t) = C_r x_r(t|t) - C\hat{x}(t+1|t) - Du_0(t-k+1) \tag{4.12}$$

where $K_f$ denotes the Kalman filter gain matrix.

Inserting Eqs. (4.10) and (4.12) into (4.11), and rearranging:

$$\hat{x}(t|t) = T_{f1}(z^{-1})\big[e(t) - z^{-1}C_r x_r(t|t)\big] + T_{f2}(z^{-1})u_0(t) \tag{4.13}$$

where

$$T_{f1}(z^{-1}) = \big[I - (I + K_f C)Az^{-1}\big]^{-1} K_f \tag{4.14}$$

$$T_{f2}(z^{-1}) = \big[I - (I + K_f C)Az^{-1}\big]^{-1}\big[K_f D + (I + K_f C)Bz^{-1}\big]z^{-k} \tag{4.15}$$

The optimal signal $u_{opt}(t)$ that optimizes the cost function $J$ can now be obtained:

$$P_c\Big\{C_r^k x_r(t|t) - CA^k\hat{x}(t|t) - \Big[CT_0(k,z^{-1})B + D\Big](W_{1k}u_{opt})(t)\Big\} + (F_{ck}u_{opt})(t) = 0 \quad (4.16)$$

The optimal condition has the following two alternative representations:

$$u_{opt}(t) = -\big\{F_{ck} - P_c\big[CT_0(k,z^{-1})B + D\big]W_{1k}\big\}^{-1}P_c\big[C_r^k x_r(t|t) - CA^k\hat{x}(t|t)\big] \quad (4.17)$$

or:

$$u_{opt}(t) = -F_{ck}^{-1}P_c\big\{C_r^k x_r(t|t) - CA^k\hat{x}(t|t) - \big[CT_0(k,z^{-1})B + D\big](W_{1k}u_{opt})(t)\big\} \quad (4.18)$$

### 4.1.2 Optimal Controller Implementation

Now observe the optimal control signal as generated from Eq. (4.17), it requires the inverse operation of the nonlinear engine model, and this is *computationally expensive and, in many cases, infeasible*. In fact, for most of the nonlinear control techniques, optimal control solutions are obtained through the inversion of the nonlinear system, and it is exactly this inversion that hurdles the application of advanced control into practical engineering. Meanwhile the optimal control signal as generated from Eq. (4.18) avoids inverse operation of the nonlinear model for its computation, thus it is ideal for real time engine control application. The block diagram for the transient engine control system and controller structure is shown in Fig. 4.7.

The two parameters $P_c$ and $F_{ck}$ are design freedom to further improve the control performance. Usually, if a PID controller exists that stabilizes the closed loop system, which is the case for most of the civil and military engine controls, experiences show that it is very easy to find a set of $P_c$ and $F_{ck}$ that makes the closed loop stable.

To summarize, the following observations are listed:

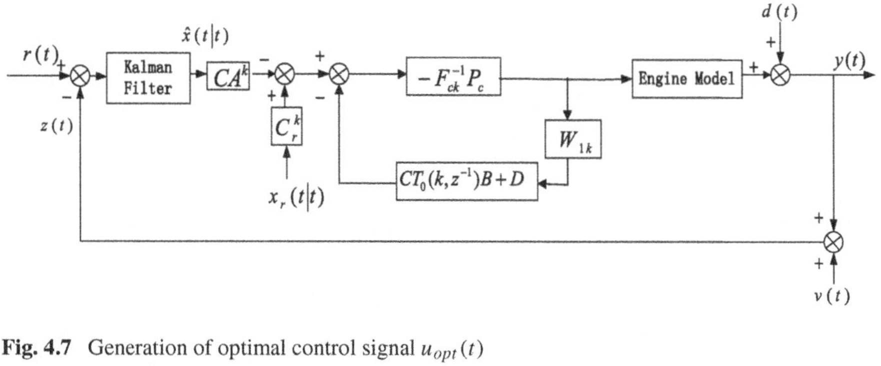


**Fig. 4.7** Generation of optimal control signal $u_{opt}(t)$

(1) Optimal control signal is provided by the state estimator, and therefore it is essentially a Kalman filter based control;
(2) There are two structures for controller implementation. As the latter one does not need the inverse operation of the nonlinear engine model, it is practically appealing for real time engine control;
(3) Two design parameters can be utilized to further improve control performance.

### *4.1.3 Fuel Flow Control of Turbofan Engines for Acceleration: Numerical Study*

Now consider the simple case for aeroengine speed closed loop control system: fuel flow control of rotational speed of the engine shaft. The following scenario is assumed: the engine is controlled by fuel flow $W_f$ and experiences acceleration from 0.9 to the design point 1.0; it is then decelerated back to 0.9. Although it is not necessary, the engine model in programming codes can be inserted into the block diagram as a non-analytical module. But to speed up the simulation process, a Hammerstein-Wiener representation is used for aeroengine model (see Fig. 4.8). The two design parameters need to be chosen: first let $F_{ck}^{-1} = -PI$ to obtain a stable closed loop control system, where *PI* denotes the PI controller of the engine; then the parameter $P_c$ is chose to be $P_c = \frac{0.5}{1-0.2z^{-1}}$. The control performance is shown in Fig. 4.9, also shown is the corresponding PI control performance.

It is seen from Fig. 4.8 that the proposed controller does not significantly improve the performance during engine acceleration and deceleration, but it is still advantageous in the following points:

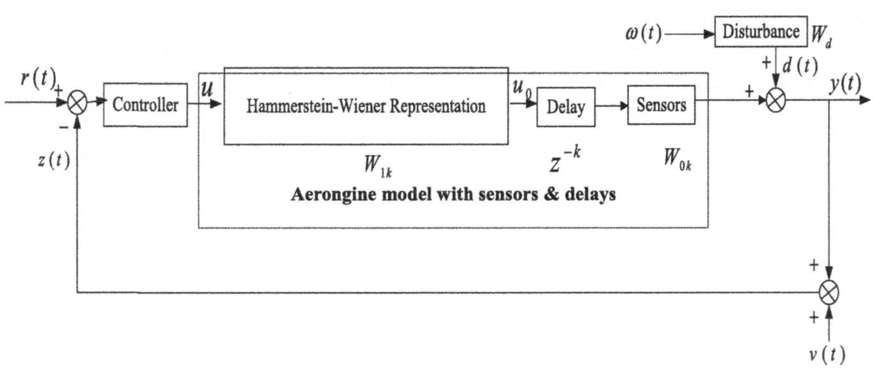


**Fig. 4.8** Block diagram of the aeroengine control system

**Fig. 4.9** Closed loop performance while engine accelerates and decelerates between 0.9 and 1.0

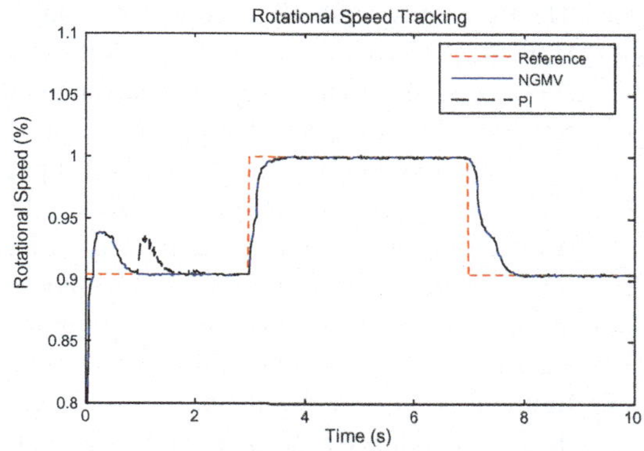


(a) Acceleration and deceleration performance of the proposed controller, as compared with that of PI control.

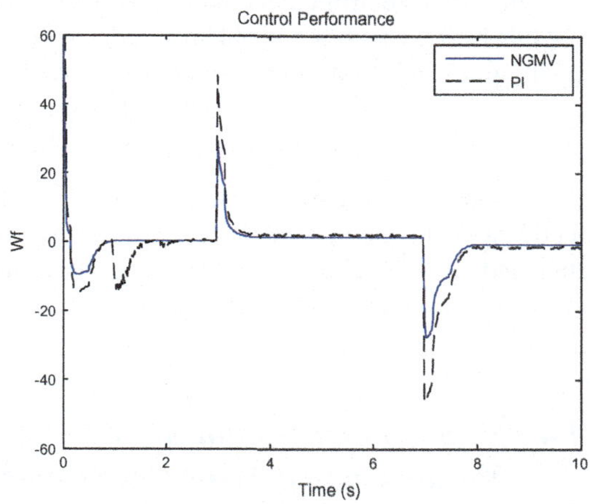


(b) Fuel flow of the proposed controller, as compared with that of PI control.

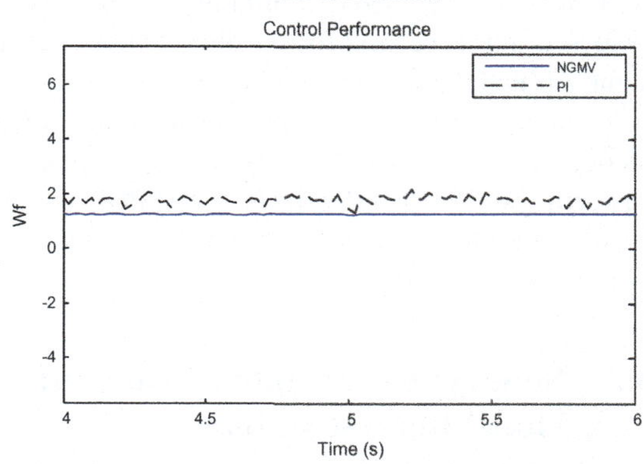


(c) Amplified fuel flow curve over 4-6 seconds.

(1) The transient overshoot is reduced over large envelope flight, and this is shown from the performance during 0–3 s. But it is warned that the observed large overshoot is not due to the PI control design but the model itself. That is, the identified model is not applicable/feasible from idle to 0.9 rotational speed of shaft. However, the proposed controller still suppresses the overshoot effectively;
(2) Fuel consumption is significantly reduced (Fig. 4.8b). Fuel consumption has been a very important performance index for both civil and military engines. It becomes increasingly important for modern engines for clean CO2 emission. In fact, a 1.5% reduction in fuel consumption index is a highly effective design.
(3) Fuel supply becomes much smooth and steady. This is shown in the amplified figure in Fig. 4.8c. Smooth oil supply can reduce the fatigue loss in fuel metering devices, and this can further cut down the fuel consumption. In fact, due to the presence of noises and disturbances in the control loop, the PI controller can amplify the detrimental effects, leading to unsteady control performance. This has been a troubling problem in engine control practice and hardware in the loop testing. In the proposed design, however, the design freedom is chosen to be $P_c = \frac{0.5}{1-0.2z^{-1}}$, and such a choice improves control performance while providing low-pass filtering effects, resulting in a suppression of noise.

Therefore, the simulation results, although still preliminary, have validated the effectiveness of the proposed design. They also provide useful guidance for practical design and lay the foundation for carrying out further research on advanced transient control and hardware in the loop testing.

### *4.1.4 Summary for Nonlinear Generalized Minimum Variance-based Aeroengine Control*

A nonlinear generalized minimum variance type of control approach has been proposed in this section for handling engine transient design problems. It has been demonstrated that this proposed method can "take in" the engine model in whatever non-analytical formats. Indeed, this approach does not consider the detailed engine model representations and is essentially an operator-based method. The simulation has demonstrated its effectiveness and this also motivates the generalization of the framework into a receding horizon setup. This is considered in the next section.

## 4.2 Nonlinear Aeroengine Transient Control with NARX Model Representation

Last chapter has seen the capability of nonlinear minimum variance-based control for handling the non-analytical issue, that is, the design methodology allows the utilization of the component-level models that may be represented by lookup tables or some other non-analytical modules such as (C or Fortran) program codes. The hurdle has thus been conquered confronting most of the control theories requiring an analytical model before control design, although a Hammerstein-Wiener representation is used for aeroengine control loop to speed up the simulation process. This bridges the gap with integrated nonlinear modelling and nonlinear control design, although the methodology is theoretically independent upon identification models. To further tell the story, other than a single-input and single-output case as in the above section, we shall develop a multivariable control case based on a structure of nonlinear autoregressive with extra input (NARX) model. This section is structured as follows: Sect. 4.2.1 presents the model description as identified as the nonlinear model with the structure of NARX; Sect. 4.2.2 utilizes the same NGMV control structure as compared with the performance of the benchmarking PID controllers, where extensive simulations are provided with different scenarios. Section 4.2.3 summarizes this nonlinear transient control with NARX model representations.

### 4.2.1 Aeroengine Modeling with NARX Representation

As introduced in Chap. 2, NARX model is understood as an extension of the linear autoregressive with extra input (ARX) structure. It combines ARX model and neural networks with multi-step prediction. NARX includes feedback connections enclosing several layers of the network. The defining equation of the NARX model is as follows (reproduced from Chap. 2):

$$y(t) = f[y(t-1), \cdots, y(t-n_a), u(t-n_k), \cdots u(t-n_k-n_b+1) \tag{2.73}$$

where: $y(t)$ is the output, $u(t)$ is the input, and $f$ is a nonlinear function or operator that depends on the information from previous inputs and outputs; $n_a$ and $n_b$ are the model orders for output and input respectively with $n_k$ denoting system delay. Thus, the NARX model consists of the regressors and the nonlinear estimator as shown in Fig. 4.10 (reproduced from Chap. 2).

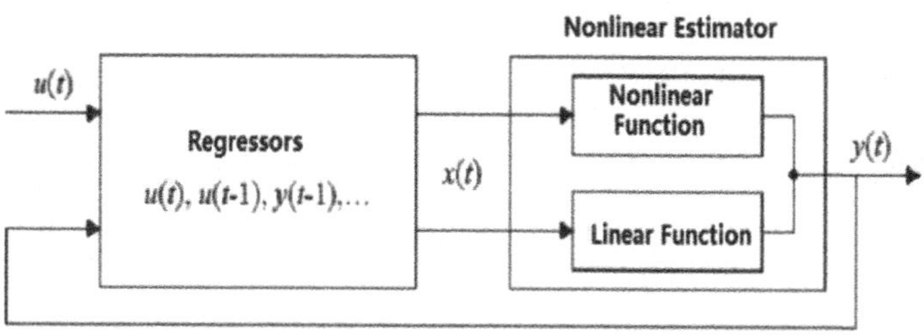


**Fig. 4.10** Typical structure of NARX model

The aeroengine inputs used in this study are the fuel flow $W_f$ and the nozzle throat area $A_8$, while the two rotational speeds $n_H$ (high-pressure) and $n_L$ (low-pressure) as outputs. The target prototype is a two-spool turbofan engine. To start with, the identification procedures as in Chap. 2 are followed. Essential steps include a well-tuned nonlinear component-level model preparing data for identification and validation (Fig. 4.11).

The multivariable counterpart of model representation as derived from (2.73) can now be written as:

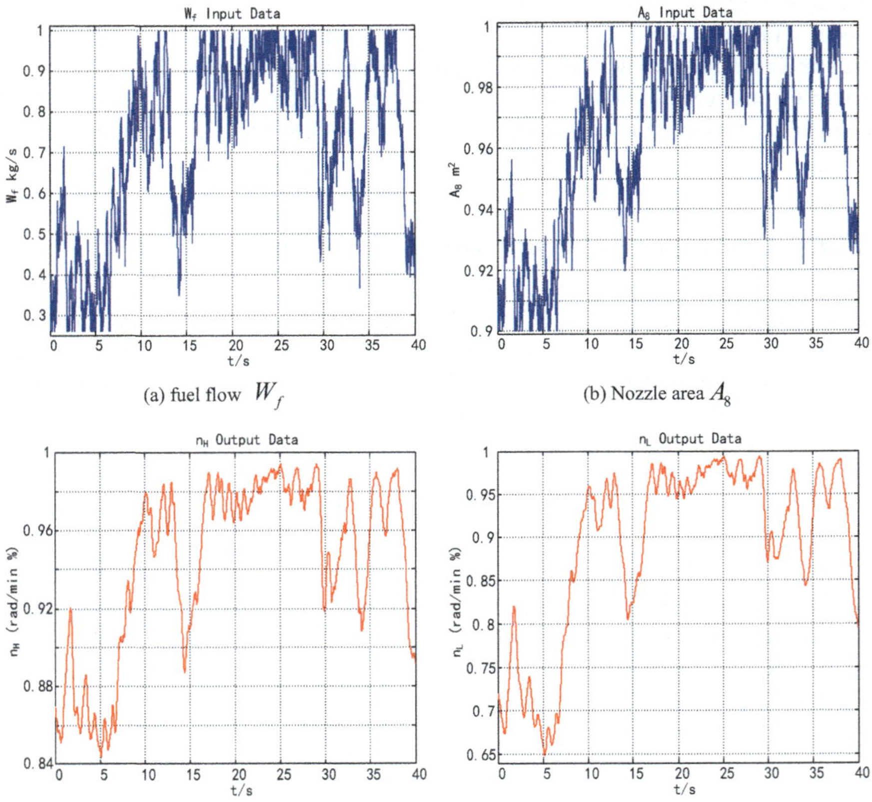


(a) fuel flow $W_f$ (b) Nozzle area $A_8$

(c) High-pressure rotational speed $n_H$ (d) Low-pressure rotational speed $n_L$

**Fig. 4.11** Input–output data for aeroengine model identification and validation

$$\begin{aligned}
n_H(t) &= [n_H(t-1), n_H(t-2), \cdots, n_H(t-n_{a11}),\\
&n_L(t-1), n_L(t-2), \cdots, n_L(t-n_{a12}),\\
&W_f(t-n_{k11}), \cdots, W_f(t-n_{k11}-n_{b11}+1),\\
&A_8(t-n_{k12}), \cdots, A_8(t-n_{k12}-n_{b12}+1)\\
n_L(t) &= [n_H(t-1), n_H(t-2), \cdots, n_H(t-n_{a21}),\\
&n_L(t-1), n_L(t-2), \cdots, n_L(t-n_{a22}),\\
&W_f(t-n_{k21}), \cdots, W_f(t-n_{k21}-n_{b21}+1),\\
&A_8(t-n_{k22}), \cdots, A_8(t-n_{k22}-n_{b22}+1)
\end{aligned} \tag{4.19}$$

where: $n_a$ is the identified model output order; $n_b$ the identified model input order; and $n_k$ is time delay, defined respectively as follows:

$$n_a = \begin{bmatrix} n_{a11} & n_{a12} \\ n_{a21} & n_{a22} \end{bmatrix}, \; n_b = \begin{bmatrix} n_{b11} & n_{b12} \\ n_{b21} & n_{b22} \end{bmatrix}; \; n_k = \begin{bmatrix} n_{k11} & n_{k12} \\ n_{k21} & n_{k22} \end{bmatrix} \tag{4.20}$$

meanwhile, to restricted the model orders by balancing the accuracy requirement, the above parameters are confined within [1,5].

Before proceed to model building, the dataset needs to be split for identification and validation. Thus, some appropriate measures are taken for estimating the model accuracy. The following indices are usually used:

$$\text{Best fit: } f_{FIT} = \left(1 - \frac{|y - \hat{y}|}{|y - \overline{y}|}\right) \times 100\% \tag{4.21}$$

where: $f_{FIT}$ is defined as model fitting degree; $y$ is the aeroengine output data (here refer to high-pressure and low-pressure rotational speeds); $\hat{y}$ is the identified NARX model output data; and $\overline{y}$ is the average of $y$.

***Akaike information criterion (AIC)***: for different model orders, calculate $f_{FPE}$ and $f_{AIC}$ defined below, and the smallest $f_{FPE}$ and $f_{AIC}$ and for the NARX model, the better accuracy of the resulting model:

$$f_{FPE} = V\left(\frac{1 + d/N}{1 - d/N}\right); \; f_{AIC} = \log V + \frac{2d}{N} \tag{4.22}$$

where: $V$ is the loss function; $\theta_N$ is the estimated parameter; $d$ is the identified parameter; and $N$ is the input–output data number. $V$ is defined below:

$$V = \det\left\{\frac{1}{N}\sum_{1}^{N} \varepsilon(t, \theta_N)[\varepsilon(t, \theta_N)]^T\right\} \tag{4.23}$$

Now multiple datasets from the nonlinear component-level model are utilized to identify the NARX model for aeroengine, and 12 models are obtained, as shown in Table 4.2, while Fig. 4.9 shows the results of the identified model, where *mi* means the model that has been obtained and *zv* is the data to validate the model. It is also

**Table 4.2** NARX models for aeroengine

| Model | $n_a$ | $n_b$ | $n_c$ | $n_H$ FIT % | $n_L$ FIT % | FPE | AIC |
|---|---|---|---|---|---|---|---|
| m1 | [1 0;0 1] | [1 1;1 1] | [1 1;1 1] | 95.7 | 90.56 | 0.0015 | −38.3193 |
| m2 | [1 0;0 1] | [1 1;1 1] | [1 2;0 2] | 95.6 | 87.08 | 0.1037 | −34.4991 |
| m3 | [1 0;0 1] | [1 2;1 1] | [1 2;0 2] | 95.32 | 87.08 | 0.1601 | −34.0636 |
| m4 | [1 0;0 1] | [1 2;1 1] | [1 2;1 1] | 95.32 | 90.59 | 0.0943 | −34.5935 |
| m5 | [1 0;0 1] | [1 3;1 1] | [1 2;1 2] | 87.96 | 88.65 | 0.1567 | −34.0846 |
| m6 | [1 1;1 1] | [1 2;1 1] | [1 2;1 1] | 92.71 | 86.20 | 0.1415 | −34.1855 |
| m7 | [2 0;0 2] | [1 0;0 1] | [1 3;3 1] | 93.85 | 78.49 | 0.0095 | −36.8857 |
| m8 | [3 0;0 3] | [1 0;0 1] | [1 4;4 1] | 90.91 | 83.59 | 0.0058 | −37.3788 |
| m9 | [4 0;0 4] | [1 2;2 1] | [1 0;0 1] | 87.24 | 80.62 | 0.0011 | −39.0354 |
| m10 | [5 0;0 5] | [1 0;0 1] | [1 0;0 1] | 85.46 | 81.71 | 0.0040 | −37.7534 |
| m11 | [1 0;0 1] | [1 1;1 1] | [1 2;1 5] | 95.63 | 86.65 | 0.0686 | −34.9117 |
| m12 | [2 1;1 2] | [2 2;2 2] | [1 2;2 1] | 78.32 | 79.69 | 0.0456 | −38.7359 |

the data collected during the similar simulation and are applied to compare with the identification model. Fit represents the model matching degree, so m1 is selected as the control plant.

To validate the model, now use the remaining dataset with the identified model $m1$ and the model accuracy is shown in Fig. 4.12. It is seen that the model error is

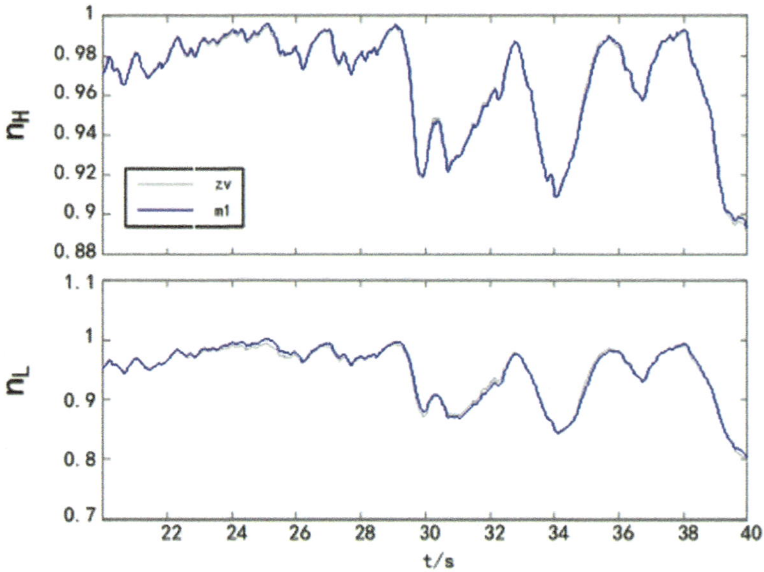


**Fig. 4.12** Model validation with best Fit NARX model m1

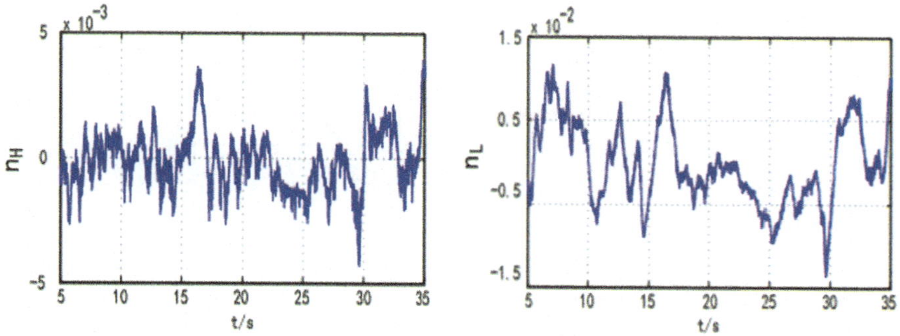


(a) Error in High-pressure Rotational Speed (b) Error in Low-pressure Rotational Speed

**Fig. 4.13** Model identification errors

relatively small for high-pressure rotational speed of 95.7%; while for low-pressure rotational speed of 90.56%.

To further clearly see the model identification accuracy, the modeling error has been illustrated in Fig. 4.13. Since the identified model is feasible for idle and above states, thus the data during first 5 s are omitted; then it is seen that for the same input, the identification error is indeed very small with high-pressure rotational speed of ±0.1%; while low-pressure rotational speed of ±0.5%. Thus the identified model is validated and can be utilized for control design investigations.

### *4.2.2 Nonlinear Aeroengine Transient Control for Benchmarking*

The simulations are carried out in this section based on the above theoretical analysis for NGMV controller design. To verify the effect of the controller, the NGMV controller and the traditional proportional–integral–derivative (PID) controller are compared. To consider the more realistic case, bounded disturbance is injected to examine the robustness of the system.

To proceed, the parameters in NGMV controller are determined, specifically $F_{ck}$ is chosen as the PID controller to ensure stability; and $P_c$ is designed to be a low-pass filter as follows:
$F_{ck}=\begin{bmatrix} 30+\frac{10}{s} & 0 \\ 0 & 10+\frac{15}{s} \end{bmatrix}$; and $P_c$ is a Butterworth filter of order-3, with a cutoff frequency of 0.97 Hz, namely:
$P_c = \begin{bmatrix} \frac{0.91s^3+2.73s^2+2.73s+0.91}{s^3+2.812s^2+2.64s+0.8281} & 0 \\ 0 & \frac{0.91s^3+2.73s^2+2.73s+0.91}{s^3+2.812s^2+2.64s+0.8281} \end{bmatrix}$. Finally, the PI controller act as: $PI = \begin{bmatrix} 60+\frac{30}{s} & 0 \\ 0 & 20+\frac{50}{s} \end{bmatrix}$.

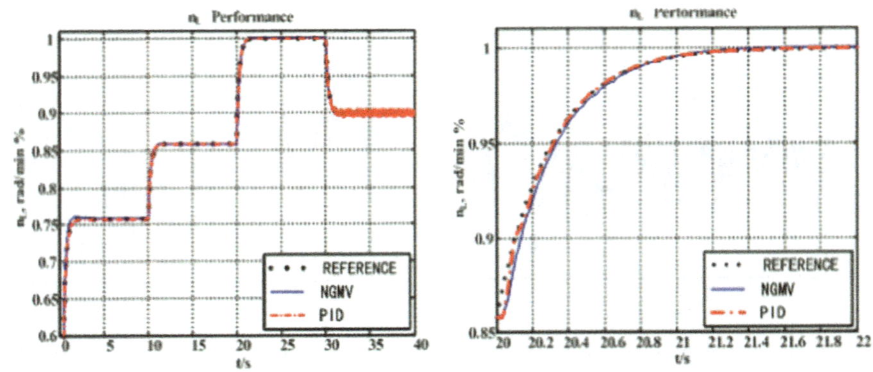


(a) Low-pressure Rotational Speed

(b) Max Low-pressure Rotational Speed

**Fig. 4.14** Comparative performance for $n_L$

### *Comparison without Disturbance*

First consider the comparative performance without disturbance, and the results are shown in Fig. 4.14 for low-pressure rotational speed $n_L$.

Figure 4.14a is the performance where the simulation time is 40 s and the rotational speed references are: 0−10 s acceleration to 76.2%; 10−20 s acceleration to 85.6%; 20−30 s acceleration to 100%; while 30−40 s deceleration to 90%. Figure 4.14b is the magnified view for the maximum power between 20 and 22 s. From (a), it is seen that both NGMV controller and PID controller can fulfill the transient control without large overshot; and from (b), the acceleration takes within 2 s which is acceptable. Meanwhile during 30−40 s, the deceleration process is also smooth. When reaching to steady state, PID control has slight oscillation and NGMV control has a marginally better performance over PID control.

Figure 4.15 is for high-pressure rotational speed $\pm 0.1$ , and similarly, the simulation time is 40 s and the rotational speed references are: 0−10 s acceleration to 76.2%; 10−20 s acceleration to 85.6%; 20−30 s acceleration to 100%; while 30−40 s deceleration to 90%. From (a), it is seen that both NGMV controller and PID controller can fulfill the transient control without large overshot; yet from 0−10 s, NGMV control does not fulfill a good tracking performance and responses sluggish over PID control, although finally achieving the tracking at around 10 s. This is due to the reason that the model is valid from idle state and does not take the starting process into account, henceforth the performance over this duration is not considered for benchmarking. From (b), it is seen that high-pressure rotational speed stabilizes at around 1.5 s with a steady-state error 0.5% while at 6 s reaching the maximum power. Meanwhile during 30−40 s, the deceleration process is also smooth. When reaching to steady state, PID control has slight oscillation and NGMV control has a marginally better performance over PID control, as confirmed by the low-pressure rotational speed.

Control performance must also refer to control signal and the fuel consumption is illustrated in Fig. 4.16. It is seen that PID control signal has a large oscillation in

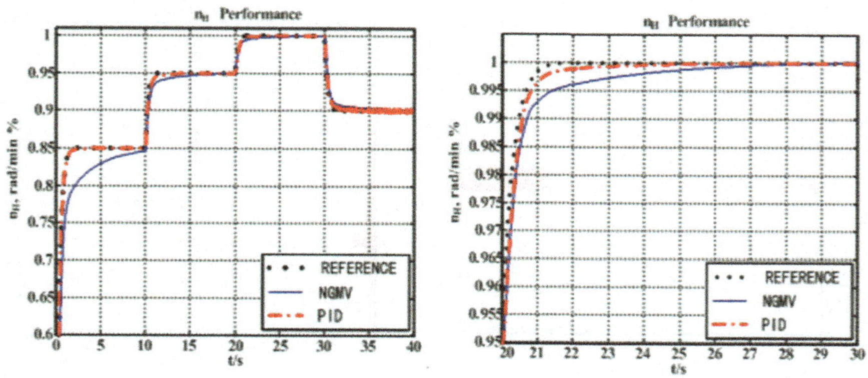


(a) High-pressure Rotational Speed

(b) Max High-pressure Rotational Speed

**Fig. 4.15** Comparative performance for $n_H$

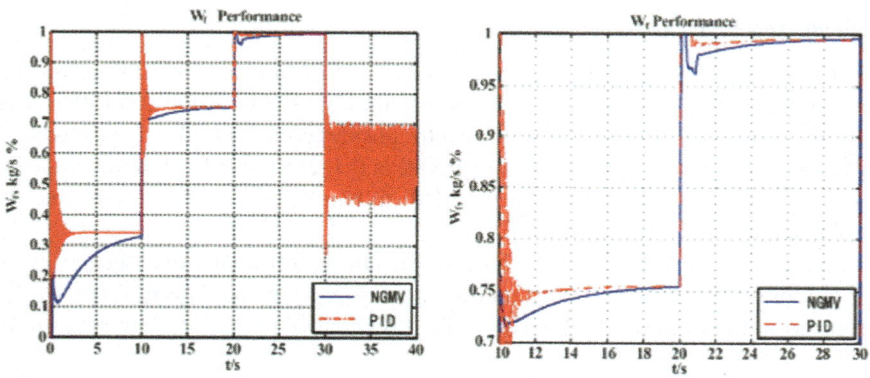


(a) Fuel Consumption

(b) Fuel Consumption for Max State

**Fig. 4.16** Comparison for fuel flow $W_f$

stark comparison with that of NGMV control. Indeed, in practice, the fuel metering devices are difficult to implement the PID control signal. For the magnified view during 10–30 s, NGMV control demonstrates a superior specific fuel consumption efficiency over that of PID control, which is very important to aeroengine economic considerations.

The fuel consumption property for both controllers are also reflected in nozzle area as shown in Fig. 4.17, where PID control causes a serious oscillation in nozzle area while NGMV control is capable of a smooth transition even over the deceleration process during 30–40 s. Indeed, PID control leads to over-excitations, which is also shown in the magnified view. This clearly demonstrates the effectiveness of the NGMV control.

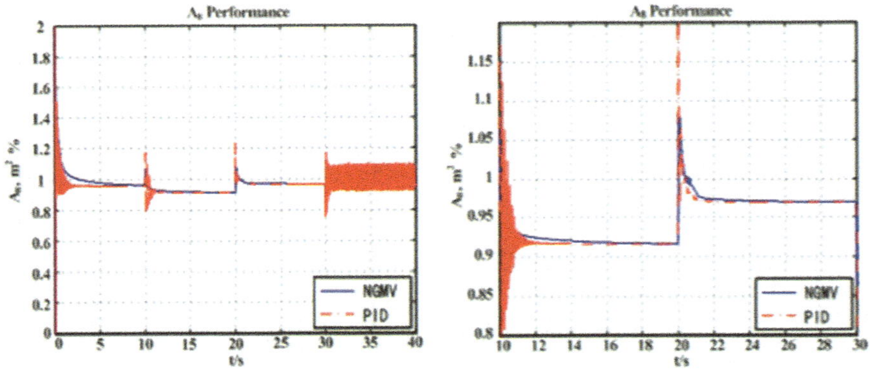


(a) Nozzle Area

(b) Nozzle Area for Max State

**Fig. 4.17** Comparison for nozzle area $A_8$

***Comparison under Disturbance***

In practice, it is unavoidable to subject a variety of disturbance and noise over the aeroengine operation. Thus, to further verify the robustness of the controller, it is pertinent to take the effect of disturbance into consideration. Since the signal has been in a corrected scale, the disturbance magnitude of 0.1 refers to 10% of the disturbance level; similarly, 0.2 refers to 20% of the disturbance level, and the frequency is designated to be 100 Hz.

Figures 4.18 and 4.19 are the performance results for difference disturbance levels for low-pressure rotational speed and high-pressure rotational speed, respectively. From (a) in both figures, both NGMV and PID control possess certain robustness

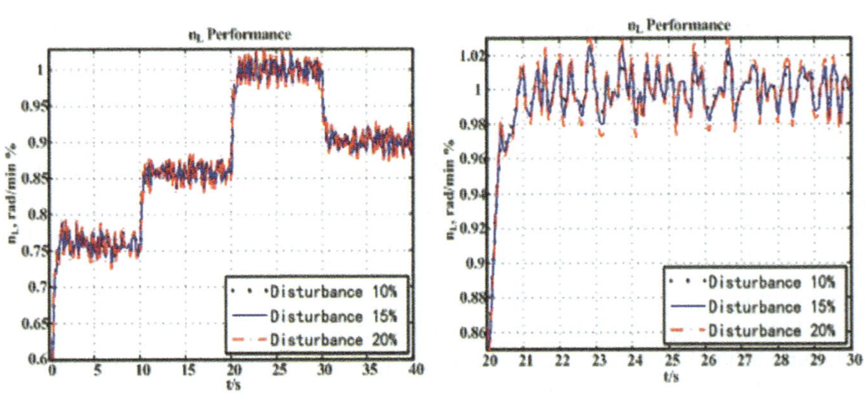


(a) Low-pressure Rotational Speed under Disturbance

(b) Max Low-pressure Rotational Speed Performance

**Fig. 4.18** Low-pressure rotational speed $n_L$ under difference disturbance levels

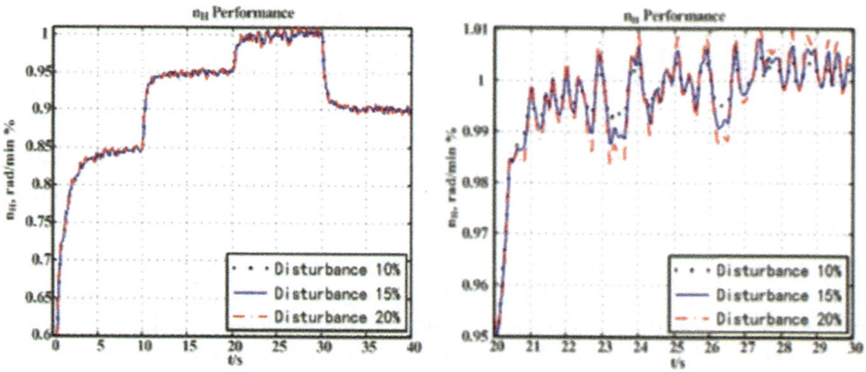


(a) High-pressure Rotational Speed under Disturbance

(b) Max High-pressure Rotational Speed Performance

**Fig. 4.19** High-pressure rotational speed $n_H$ under difference disturbance levels

against disturbances. When the disturbance level does not exceed 15%, the steady-state error in both high-pressure rotational speed and low-pressure rotational speed is below 1%. Also, it is well-known that the design point performance is one of the most concerned issue: now the steady-state error is less than 0.5% with a disturbance level up to 15%, and this satisfies the control system requirement. However, when the disturbance level reaches to 20%, the steady-state error in low-pressure rotational speed increases to 2%; and high-pressure rotational speed steady-state error also approaches around 1%. This would not conform to the control system requirement. It is thus concluded that NGMV control is indeed robust against disturbances even up to 15%, which has been a significant level.

### 4.2.3 *Summary for NGMV Control of NARX Model Representation*

This section has introduced the NGMV formalism as applied for turbofan engine represented by an NARX model structure. This type of model suits perfectly the situation where it is time-consuming to model the system dynamics. The proposed method demonstrates that its fuel consumption is more efficient with a smooth modulation than the traditional PID-type controllers. Meanwhile, the method also possesses robustness to disturbances which may upset conventional design. Since the approach is designed for multivariable control systems, variable cycle engine even adaptive engine control system can be further studied by adopting this framework for future investigations.

## 4.3 Nonlinear Predictive Generalized Minimum Variance-Based Aeroengine Transient Control

Model based Predictive Control (MBPC) has been very successful in industrial applications over the past three decades. The most popular predictive control algorithms are Dynamic Matrix Control (DMC) [28, 29] and Generalized Predictive Control (GPC) [30, 31]. The GPC approach is not very suitable for multivariable constrained systems commonly encountered in the process industries and as a consequence, almost all vendors have adopted a DMC-like approach [32, 33]. However, these approaches are based on linear models and are thus inadequate to handle systems described by nonlinear models. Moreover, most systems are inherently nonlinear and this necessitates developing methods utilizing nonlinear models to achieve optimal performance over wide range of system operations.

There has been a rich history of research in the field of nonlinear model predictive control (NMPC). Many approaches have been proposed such as infinite horizon NMPC [34] and quasi-infinite horizon NMPC [35], contractive NMPC [36] etc. Excellent reviews of existing NMPC techniques can be found in [37–39]. Commercially available NMPC technology can be found in the recent article [40].

One of the key questions in NMPC is closed-loop stability. As a result, most of the approaches employ a Lyapunov-based strategy where construction of Lyapunov functions for stability guarantee, e.g. via the introduction of end constraints becomes the main concern. Unlike these "stability guaranteed" approaches, the control strategy introduced here is motivated by the need to produce a control law, which is very simple to implement in industry [41]. In fact, many of the stability-guaranteed NMPC are difficult to implement [36] and there is a natural demand for a practical control law for nonlinear systems that has a sound but relatively simple theoretical basis. A promising approach is based on the generalization of the nonlinear generalized minimum variance control that has been developed in Sect. 4.1. The new development here is an extension to time-varying systems as well as the new feature of handling input saturation constraint over previous results [42–44]. The presentation of this paper is as follows: the nonlinear system model is introduced in Sect. 4.3.1; the criteria of determining performance index is discussed in Sect. 4.3.2; the optimal predictive control law and its implementation is obtained in Sects. 4.3.3, 4.3.4 addresses handling of input saturation; Sect. 4.3.5 provides a case study to demonstrate the proposed design strategy and finally Sect. 4.3.6 concludes the work.

### *4.3.1 Aeroengine Representation and Signal Listing*

The aeroengine transient control system is illustrated in Fig. 4.20 in the Sect. 4.1. It is reproduced here for ease of reference.

To proceed, the blocks are also replicated here:

**Non-analytical module** can be expressed as:

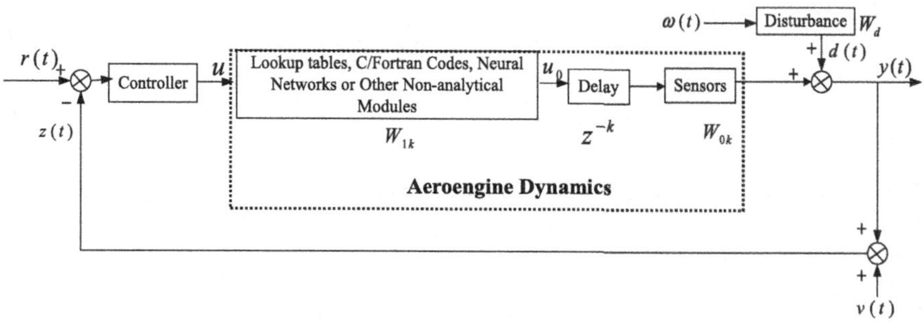


**Fig. 4.20** Block diagram for transient control of aeroengines

$$u_0 = (W_{1k}u)(t) \tag{4.24}$$

**Linear Subsystem with Channel Delay**: The linear subsystem including the channel delay is represented by the following linear time-invariant state-space model:

$$\begin{aligned} x_0(t+1) &= A_0 x_0(t) + B_0 u_0(t-k) + E_0 \varphi(t) \\ y_0(t) &= C_0 x_0(t) \end{aligned} \tag{4.25}$$

where: $\varphi(t)$, together with $\omega(t)$ that drives the disturbance model to be introduced below, is assumed to be a zero-mean white noise signal with identity covariance matrices. The linear disturbance model is written down in state-space form as follows:

$$\begin{aligned} x_d(t+1) &= A_d x_d(t) + B_d \omega(t) \\ d(t) &= C_d x_d(t) \end{aligned} \tag{4.26}$$

This disturbance model can then be combined with the above linear subsystem:

$$\begin{aligned} x(t+1) &= Ax(t) + Bu_0(t-k) + E\xi(t) \\ y(t) &= Cx(t) + Du_0(t-k) \end{aligned} \tag{4.27}$$

with: $A = \begin{bmatrix} A_0 & O \\ O & A_d \end{bmatrix}$, $B = \begin{bmatrix} B_0 \\ 0 \end{bmatrix}$, $E = \begin{bmatrix} E_0 & O \\ O & B_d \end{bmatrix}$, $C = \begin{bmatrix} C_0 & C_d \end{bmatrix}$ and $D = D_0$.

The vector signals are: $x(t) = \begin{bmatrix} x_0(t) \\ x_d(t) \end{bmatrix}$ and $\xi(t) = \begin{bmatrix} \phi(t) \\ \omega(t) \end{bmatrix}$.

The signals shown in Fig. 4.20 can now be listed as follows:

$$\textbf{Error signal} : e(t) = r(t) - z(t) \tag{4.28}$$

$$\textbf{Plant output} : y(t) = d(t) + (Wu)(t) \tag{4.29}$$

$$\text{with } (Wu)(t) = z^{-k} W_{0k}(W_{1k}u)(t) \tag{4.30}$$

$$\textbf{Observation signal}: z(t) = y(t) + v(t) \tag{4.31}$$

The future values of states and outputs can also be obtained as follows:

$$\begin{aligned} x(t+k+i) &= A_{k+i}^{i} x(t+k) \\ &+ \sum_{j=0}^{i-1} A_{k+i}^{i-j-1}[B(t+k+j)u_0(t+j) + E(t+k+j)\xi(t+k+j)] \end{aligned} \tag{4.32a}$$

$$y(t+k+j) = C(t+k+j)x(t+k+j) + D(t+k+j)u_0(t+i) \tag{4.32b}$$

where the algebra of A is defined as:

$$A_m^n = A(t+m-1)A(t+m-2)\cdots A(t+m-n) \text{ with } m \geq n \text{ and } A_m^0 = I.$$

## 4.3.2 *Aeroengine Performance Index Selection and Optimization*

The nonlinear optimal control laws often involve the "inversion" of the nonlinear process, however, the non-analytical module in Fig. 4.20 may prevents such an inversion operation and consequently, a mechanism must be devised to obtain the optimal control law. The performance index is carefully chosen so that a simple control law is produced. The other criterion is that the resulting control law should recover the conventional Generalized Predictive Control (GPC) controller, when the nonlinear subsystem becomes linear. This is very important since the equivalence will provide considerable confidence in generalizing the control law to nonlinear systems.

To motivate the choice of the performance index to be minimized in the nonlinear predictive GMV framework, a brief review of the derivation of the GPC controller is provided where for the moment the input will be taken to be that for the linear time varying subsystem. In GPC, the performance index to be minimized may be defined as:

$$J = E\left\{\sum_{i=0}^{N} e(t+k+i)^T e(t+k+i) + \sum_{j=0}^{N_u} \lambda_i^2 u_0(t+j)^T u_0(t+j)|t)\right\} \tag{4.33}$$

where: $E\{\bullet|t\}$ is the conditional expectation operator, conditioned on measurements up to time t$t$; $N$ is the prediction horizon, $k$ is the process delay and $N_u$ is the control horizon. $\lambda_j$ denotes the control signal weighting. Using the notations introduced above the performance index (4.33) can be rewritten as:

$$J = E\{\Psi^T\Psi + U_0^T \Psi U_0 | t\} \tag{4.34}$$

where: $\Psi = \left[e(t+k) \cdots e(t+k+N)\right]^T$ and $\Lambda^2 =$ diagonal$\{\lambda_0^2\ \lambda_1^2 \cdots \lambda_{Nu}^2\}^T$.

To minimize $J$ the procedure for optimizing the deterministic signals can be utilized and this results in setting the gradient of the cost function (with respect to control $U_0$) to zero. Then the vector of GPC control law can be shown to be:

$$U_0 = \left(V_{NN}^T V_{NN} + \Lambda^2\right)^{-1} V_{NN}^T [R - C_N \overline{x}(t+k|t)] \tag{4.35}$$

where the signal vectors are:

$R = \begin{bmatrix} r(t+k) \\ \vdots \\ r(t+k+N) \end{bmatrix}$ and $U_0 = \begin{bmatrix} u_0(t) \\ \vdots \\ u_0(t+N) \end{bmatrix}$; $C_N = \begin{bmatrix} C(t+k) \\ C(t+k+1)A(t+k) \\ C(t+k+2)A_{k+2}^2 \\ \vdots \\ C(t+k+N)A_{k+N}^N \end{bmatrix}$ with the matrix $V_{NN}$ as:

$$V_{NN} = \begin{bmatrix} D(t+k) & 0 & 0 & \cdots & 0 & \\ C(t+k+1)B(t+k) & D(t+k+1) & 0 & \cdots & 0 & \\ C(t+k+2)A(t+k+1)B(t+k) & C(t+k+2)B(t+k+1) & 0 & \cdots & 0 & \\ \vdots & \vdots & \vdots & \ddots & & \vdots \\ C(t+k+N)A_{k+N}^{N-1}B(t+k) & C(t+k+N)A_{k+N}^{N-2}B(t+k) & \cdots & C(t+k+N)B(t+k+N-1) & & D(t+k+N) \end{bmatrix}$$

In deriving (4.35), the output has been assumed to be that of the combined linear model. Now (4.35) can be rewritten as:

$$V_{NN}^T [R - C_N \overline{x}(t+k|t) - V_{NN} U_0] - \Lambda^2 U_0 = 0 \tag{4.36}$$

That is:

$$V_{NN}^T \Psi - \Lambda^2 U_0 = 0 \tag{4.37}$$

Consider now a new signal to be minimized involving a weighted sum of error and control input as follows:

$$\phi(t) = P_c e(t) + (F_{c0} u_0(t-k))(t) \tag{4.38}$$

The vector of future values of the signal may be written as:

$$\Phi = P_{cN}\Psi + F_{c0N}U_0 \tag{4.39}$$

Comparing with (4.37), it is seen that if the weightings in the above equation are defined as:

$$P_{cN} = V_{NN}^T \tag{4.40}$$

and

$$F_{c0N} = -\Lambda^2 \tag{4.41}$$

then minimization of the performance index (4.33) will be fully equivalent to the minimization of the following generalized minimum variance (GMV) type of performance index:

$$J = E\{\Phi^T\Phi|t\} \tag{4.42}$$

with $\Phi = \left[\phi(t+k) \cdots \phi(t+k+N)\right]^T$ as defined in (4.38) and (4.39). This is the case since the minimization of the GMV index (4.42) results in setting $\Phi$ to be zero and (4.39) then becomes (4.37).

These results can be summarized as follows:

**Proposition (Connection with GPC)** *The minimization of the GPC performance index (4.33) is equivalent to the minimization of the GMV form of performance index defined by*

$$J = E\{\Phi^T\Phi|t\} \tag{4.42}$$

*with:* $\Phi = P_{cN}\Psi + F_{c0N}U_0$*and the weightings defined by:*

$$P_{cN} = V_{NN}^T \tag{4.40}$$

*and*

$$F_{c0N} = -\Lambda^2 \tag{4.41}$$

***Proof*** The proof follows easily from the above reason. □

As a consequence of this connection, the minimization of the GMV performance index (4.42) will result in the same control law with GPC, if the system dynamics is linear. This provides the confidence in choosing (4.42) as the performance index for the nonlinear predictive GMV control problem. However, from Fig. 4.20 it is seen that the actual input to the system is the control signal $u$(t), rather than the input to the linear subsystem. A new signal may then be defined including the weighting of the control signal $u_0$ as follows:

$$\phi(t) = P_c e(t) + (F_{c0}u_0(t-k))(t) + F_{ck}u(t-k) \tag{4.43}$$

The vector of future values of the signal can be written as:

$$\Phi = P_{cN}\Psi + F_{c0N}U_0 + F_{ckN}U \tag{4.44}$$

where: $U = \left[\, u(t) \ \cdots \ u(t+N) \,\right]^T$ and weightings $P_{cN} = V_{NN}^T$ and $F_{c0N} = -\Lambda^2$ as in (4.40) and (4.41); the weighting $F_{ckN}$ will have simple diagonal form:

$$F_{ckN}U = diag\big( (F_{ck}u)(t) \ \cdots \ (F_{ck}u)(t+N) \big) \tag{4.45}$$

The performance index involves the minimization of the variance of (4.44) defined as follows:

$$J = E\{\Phi^T \Phi | t\} \tag{4.46}$$

This performance index will be adopted as that of the nonlinear predictive GMV control problem. As having indicated in the above reasoning: if the nonlinear subsystem in Fig. 4.20 is absent, the control costing term $F_{ckN}$ can be set to zero. Then the minimization of (4.46) is equivalent to the minimization of (4.33) according to the above Proposition; the resulting optimal controller then reverts to the conventional GPC controller. This is the fundamental rationale behind selecting (4.46) as the performance index to the nonlinear predictive GMV control problem.

### *4.3.3 Aeroengine Optimal Control Law Design & Implementation*

The optimal control law comes from setting (4.44) to zero:

$$P_{cN}\Psi + F_{c0N}U_0 + F_{ckN}U = 0 \tag{4.47}$$

That is:

$$V_{NN}^T[R - C_N\overline{x}(t+k|t) - V_{NN}U_0] - \Lambda^2 U_0 + F_{ckN}U = 0 \tag{4.48}$$

Since $U_0 = \left[\, (W_{1k}u)(t) \ \cdots \ (W_{1k}u)(t+N) \,\right]^T = W_{1kN}U$ then (4.48) can be rewritten as:

$$V_{NN}^T[R - C_N\overline{x}(t+k|t)] + \left[F_{ckN} - (V_{NN}^T V_{NN} + \Lambda^2)W_{1kN}\right]U = 0 \tag{4.49}$$

The optimal predicted state $\overline{x}(t+k|t)$ can be computed from (4.32a–b) as:

$$\overline{x}(t+k|t) = A_k^k \overline{x}(t|t) + T\left(k,\, z^{-1}\right)u_0(t) \tag{4.50}$$

where: $T\left(k,\, z^{-1}\right) = \sum_{j=0}^{k-1} A_k^{k-1-j} B(t+j) z^{-(k-j)}$.

The state $\overline{x}(t|t)$ can again be computed using the well-known time-varying Kalman filter.

Now inserting (4.50) into (4.49) results in the following optimality condition:

$$\begin{aligned} &V_{NN}^T\left[R - C_N A_k^k \overline{x}(t|t)\right] \\ &\quad + \left\{F_{ckN} - [V_{NN}^T V_{NN} + \Lambda^2 + V_{NN}^T C_N T\left(k,\, z^{-1}\right) C_{I0}] W_{1kN}\right\} U = 0 \end{aligned} \tag{4.51}$$

with: $C_{I0} = \left[I\ 0 \cdots 0\right]$.

The optimal control law can then be obtained in the following two alternative forms:

$$\begin{aligned} U_{opt} &= -\left[F_{ckN} - \left(V_{NN}^T V_{NN} + \Lambda^2 + V_{NN}^T C_N T\left(k,\, z^{-1}\right) C_{I0}\right) W_{1kN}\right]^{-1} \\ &V_{NN}^T\left[R - C_N A_k^k \overline{x}(t|t)\right] \end{aligned} \tag{4.52}$$

or:

$$\begin{aligned} U_{opt} &= -F_{ckN}^{-1}\left\{V_{NN}^T\left[R - C_N A_k^k \overline{x}(t|t)\right]\right. \\ &\quad \left. - \left[\left(V_{NN}^T V_{NN} + \Lambda^2 + V_{NN}^T C_N T\left(k,\, z^{-1}\right) C_{I0}\right) W_{1kN}\right] U\right\} \end{aligned} \tag{4.53}$$

These important results are summarized in the following statement:

**Theorem: Nonlinear Predictive GMV (NPGMV) Optimal Control Solution** *Consider the nonlinear delayed multivariable system represented by Eqs. (4.24–4.26) as in Fig. 4.20, the control law that minimizes the multi-step predictive cost function.*

$$\begin{aligned} J &= E\left\{\sum_{i=0}^{N} \phi(t+k+i)^T \varphi(t+k+i)|t)\right\} \\ &= E\left\{\Phi^T \Phi|t)\right\} = E\left\{trace\left(\Phi\Phi^T\right)|t)\right\} \end{aligned}$$

with $\Phi \equiv P_{cN}\Psi + F_{c0N} U_0 + F_{ckN} U$ is given by:

$$\begin{aligned} U_{opt} &= -\left[F_{ckN} - \left(P_{cN} P_{cN}^T - F_{c0N} + P_{cN} C_N T\left(k,\, z^{-1}\right) C_{I0}\right) W_{1kN}\right]^{-1} \\ &P_{cN}\left[R - C_N A_k^k \overline{x}(t|t)\right] \end{aligned} \tag{4.54}$$

where: $P_{cN} = V_{NN}^T$ and $F_{c0N} = -\Lambda^2$. The current control can be more conveniently obtained from the first component in the alternative expression:

$$\begin{aligned} U_{opt} &= -F_{ckN}^{-1}\left\{P_{cN}\left[R - C_N A_k^k \overline{x}(t|t)\right]\right. \\ &\quad \left. - \left[\left(P_{cN} P_{cN}^T - F_{c0N} + P_{cN} C_N T\left(k,\, z^{-1}\right) C_{I0}\right) W_{1kN}\right] U\right\} \end{aligned} \tag{4.55}$$

***Proof*** The proof follows from collecting the results leading to the controllers (4.53) and (4.54). Replacing $P_{cN} = V_{NN}^T$ and $F_{cON} = -\Lambda^2$ gives the desired result. □

It is noted that the optimal control law (4.54) is of more conceptual interest and the second form of the controller (4.55) will be of great value for implementation. To proceed, now invoking the receding horizon principle, the optimal control signal $u$(t) can be obtained by left-multiplying $C_{I0} = \begin{bmatrix} I & 0 & \cdots & 0 \end{bmatrix}$ on both sides of (4.55):

$$\begin{aligned} u_{opt} &= -C_{I0}F_{ckN}^{-1}\{P_{cN}[R - C_N A_k^k \overline{x}(t|t)] \\ &\quad - [(P_{cN}P_{cN}^T - F_{cON} + P_{cN}C_N T(k, z^{-1})C_{I0})W_{1kN}]U\} \end{aligned} \tag{4.56}$$

Or:

$$\begin{aligned} u_{opt} &= -C_{I0}F_{ckN}^{-1}\{P_{cN}[R - C_N A_k^k \overline{x}(t|t)] \\ &\quad - (P_{cN}P_{cN}^T + \Lambda^2)W_{1kN}U - P_{cN}C_N T(k, z^{-1})W_{1k}u(t)\} \end{aligned} \tag{4.57}$$

Decompose the symmetric matrix $(P_{cN}P_{cN}^T + \Lambda^2)$ into $\begin{bmatrix} Y_1 & Y_2 \end{bmatrix}$ such that:

$$(P_{cN}P_{cN}^T + \Lambda^2)W_{1kN}U = Y_1 W_{1k}u(t) + Y_2 W_{1k(N-1)}U^f \tag{4.58}$$

where $U^f$ is the vector of future control signals and $W_{1k(N-1)}U^f = \begin{bmatrix} W_{1k}u(t+1) & \cdots & W_{1k}u(t+N) \end{bmatrix}$.

Substitute (4.58) into (4.57) the optimal control signal can now be expressed as:

$$\begin{aligned} u_{opt} &= -C_{I0}F_{ckN}^{-1}\{P_{cN}[R - C_N A_k^k \overline{x}(t|t)] \\ &\quad - Y_2 W_{1k(N-1)}U^f - [Y_1 + P_{cN}C_N T(k, z^{-1})]W_{1k}u(t)\} \end{aligned} \tag{4.59}$$

Noting $U^f = C_{0I}U, C_{0I} = \begin{bmatrix} 0 & I \end{bmatrix}$, the optimal control signal (4.59) can be conveniently implemented as in Fig. 4.21.

**Remark 1** It is noted from Fig. 4.21 that the implementation of the optimal control signal $u$(t) does not rely on the inverse of the nonlinear system, thus reducing the complexity and computational burden of the controller usually associated with nonlinear controls. This provides confidence in producing a simple control law that is easy to implement in industrial applications.

### *4.3.4 Constraint Handling, Robustness and Small Control*

One of the most important constraints for any control system design is the input constraint, e.g. actuators will eventually saturate. This can cause detrimental effects on system performance. The power of the NPGMV control is that it can handle the

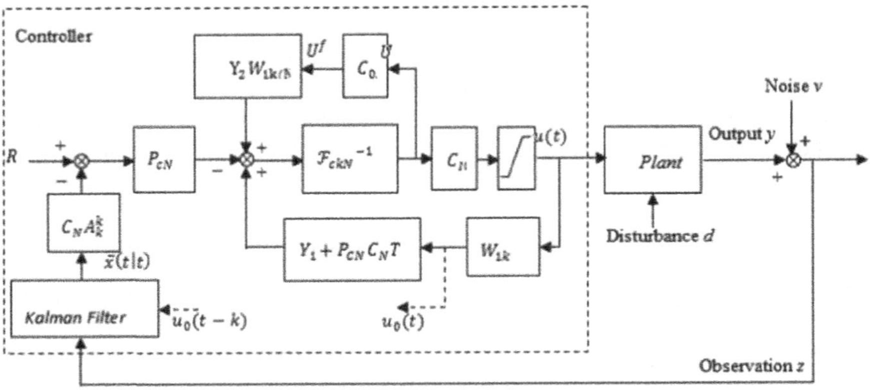


**Fig. 4.21** Nonlinear predictive GMV optimal controller implementation with input saturation

input saturation easily. This is so since the input saturation characteristic $Sat(\cdot)$ can be viewed as a nonlinear function of the input signal $u$(t). This nonlinearity can be included into the nonlinear module $W_{1k}$:

$$W_{1k}^{S} = [(W_{1k}Sat)u](t) \tag{4.60}$$

Then all the results derived above are valid with respect to the nonlinear subsystem $W_{1k}^{S}$. In specific, the optimal control signal (4.59) holds for $W_{1k}^{S}$. Then the implementation scheme still applies and is clearly equivalent to simply applying the saturation characteristics to the optimal control signal $u$(t) as illustrated in Fig. 4.21. Therefore, handling input saturation may only involve detuning the weighting matrices $F_{ck}$. No additional controller and computational complexities are required.

**Remark 2** Extensive simulations have shown that the input saturation can be handled even without detuning the weighting matrices. But it should be noted that in practice, performance may deteriorate due to the fact that saturation can have changing dynamics, and this may introduce fluctuations to control signals.

Meanwhile, for any control system, it is important to be robust against modeling uncertainties. Comparing with mathematically analytical models, the proposed system structure regarding $W_{1k}$ as a non-analytical module may have the potential to be more accurate. But an analysis of the influence of a modeling error in $W_{1k}$ on control performance is still necessary. However, as the optimal control signal involves future predictions that are computed through $W_{1k}$, and the number of computations using $W_{1k}$ is equal to the prediction horizon $N$. The analysis can be very complicated and rule-of-thumb design guidelines may be appropriate.

As the computational burden can be relieved using a small prediction horizon, e.g. $N = 1$ is adequate. For real time applications, a small prediction horizon $N$ and a small control weighting $F_{c0N} = -\Lambda^2$ are applied (see the discussion below Fig. 4.22 in the next section). In this case, controller performance behaves similar to

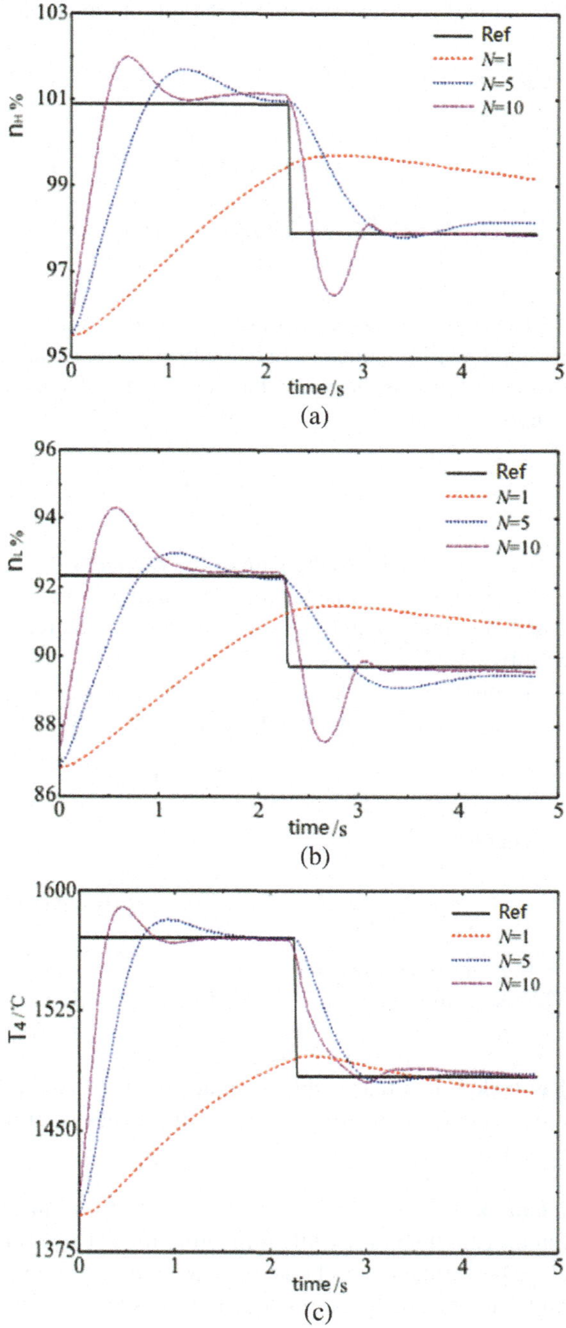


**Fig. 4.22** Aeroengine performance for different $N$: **a** High-pressure rotational speed; **b** Low-pressure rotational speed; and **c** Turbine temperature

the following nonlinear generalized minimum variance control:

$$u_{opt}(t) = -F_{ck}^{-1} P_c\big(C_r^k x_r(t|t) - CA^k\hat{x}(t|t) - \big[CT_0(k, z^{-1})B + D\big]\big(W_{1k}u_{opt}\big)(t)\big\} \tag{4.61}$$

This can be further manipulated into:

$$u_{opt}(t) = -F_{ck}^{-1} T_1(z^{-1})P_c[r(t) - d(t)] + F_{ck}^{-1}\big[T_1(z^{-1})P_c W_0 - T_2(z^{-1})\big]u_0(t) \tag{4.62}$$

The above equation represents a feedback loop, and with the assumption that the nonlinear module, the linear subsystem is $L$-stable, an application of small gain theorem results in the claim that the system will closed-loop $L$-stable if the following condition is satisfied:

$$\gamma\big(F_{ck}^{-1}\big[T_1(z^{-1})P_c W_0 - T_2(z^{-1})\big]\big) < \frac{1}{\gamma(W_{1k})} \tag{4.63}$$

where $\gamma(\cdot)$ denotes the gain of the $L$ operator.

It remains to see that a choice of small gain in $F_{ck}^{-1}$ will decrease the closed-loop gain. This implies that system robustness is increased (due to small gain theorem again).

To proceed, now define $W_{1k}^p = W_{1k}^n + \Delta$ with $W_{1k}^p$ the real plant, $W_{1k}^n$ the model and $\Delta$ the additive uncertainty. The output $y^p$ now becomes:

$$y^p = d + W_{0k}\big[\big(W_{1k}^n + \Delta\big)(u)\big] \tag{4.64}$$

That is:

$$y^p = y^n + W_{0k}\Delta(u) \tag{4.65}$$

If $\Delta(u)$ decreases with $u$, then $(y^p - y^n)$ will decrease for large gain in $F_{ck}$. That is, a small gain in $F_{ck}^{-1}$ will reduce the value in $(y^p - y^n)$ while enhancing the robustness.

**Remark 3** It is reasonable to assume that the system uncertainty is associated with the nonlinear module $W_{1k}$ while the linear subsystem $W_{ok}$ can be relatively accurately modeled.

**Remark 4** It is seen that small gain in $F_{ck}^{-1}$ can enhance the system robustness (small gain theorem) while mitigating the detrimental effect of additive uncertainty $\Delta$ on performance. A choice of small control signal is also desirable as for practical applications, small control signal implies less energy consumption.

### 4.3.5 Nonlinear Predictive GMV-Based Aeroengine Transient Control

Application of the nonlinear predictive GMV-based control design can be applied to aeroengine transient control. It is seen that the three parameters $F_{ckN}$, -$\Lambda^2$, and $N$ need to be determined. First of all, $F_{ckN}$ is usually chosen as of the PID control form, similar with GMV controller. That is, an PID controller is first designed which provides satisfactory performance, e.g. $C_{PID}(z^{-1})$; then simply $F_{ckN} = C_{PID}(z^{-1})$ for parameter $F_{ckN}$.

The parameter $\Lambda^2$ represents the weighting in control signal. If the elements are small, this implies that the designer wishes to take penalty on performance while accepting a large control power consumption; similarly, large elements in $\Lambda^2$ indicates a penalty on control signal. Usually, it is chosen as $\Lambda^2 = \lambda \times diag(a, b, c)$, and in the following numerical example, $a$, $b$, and $c$ are 1, 1, and 0.6, respectively. Then $\Lambda^2$ will be solely determined by $\lambda$, and its effect on control performance will be evaluated subsequently.

Finally, the parameter $N$ is the prediction horizon, whose values determine the number of nonlinear subsystems to be deployed in the controller implementation. Its effect is also to be evaluated. In summary, two key parameters $\lambda$ and $N$ need to be determined. And it is of great interest to provide evaluation to their effects.

Now the control performance for different $N$ is presented in Fig. 4.22. It is seen that the smaller the parameter $N$, the larger the system damping, leading to sluggish system response; conversely, a large parameter $N$ leads to decreased system damping, henceforth resulting in large overshot. Henceforth, if aeroengine transient performance is concerned, it is preferred for a large $N$, yet bearing in mind that a large $N$ also requires substantial computational resource.

Now for given $N$, it is of interest to evaluate aeroengine performance for different $\lambda$. To start with, set $N = 1$ for fast computation, and the effect of $\lambda$ parameter can be evaluated as shown in Fig. 4.23. It is seen that a small $\lambda$ does result in a small system damping, henceforth a large overshot. Thus, for allowable $N$, it is suggested to chose a small $\lambda$ for fast transition.

From the above investigation, it is also noticed that a simultaneous decreasing in $\lambda$ and increasing in $N$ can surely provide good transient performance, yet might also lead to large overshot. In practice, safety and reliability are also concerned; henceforth both $\lambda$ and $N$ must be constrained. A systematic evaluation must be conducted for determination of $\lambda$ and $N$ for practical design of aeroengine transient control. A numerical example can also be provided to validate the claim as shown in Figs. 4.24 and 4.25, where $\lambda = 0.08$ and $N = 7$ are prescribed and compared with the corresponding PID control performance. It is seen clearly that PID control has a quick response yet a large overshoot in comparison with NPGMV. Usually, control effort will be panelized, should a control lead to fast response. Yet, Fig. 4.24, NPGMV controller has a smooth control in three control variables in comparison with those from PID controller. For the current aeroengine discussion, this will imply that NPGMV control can benefit for fuel metering devices and thus is preferred for

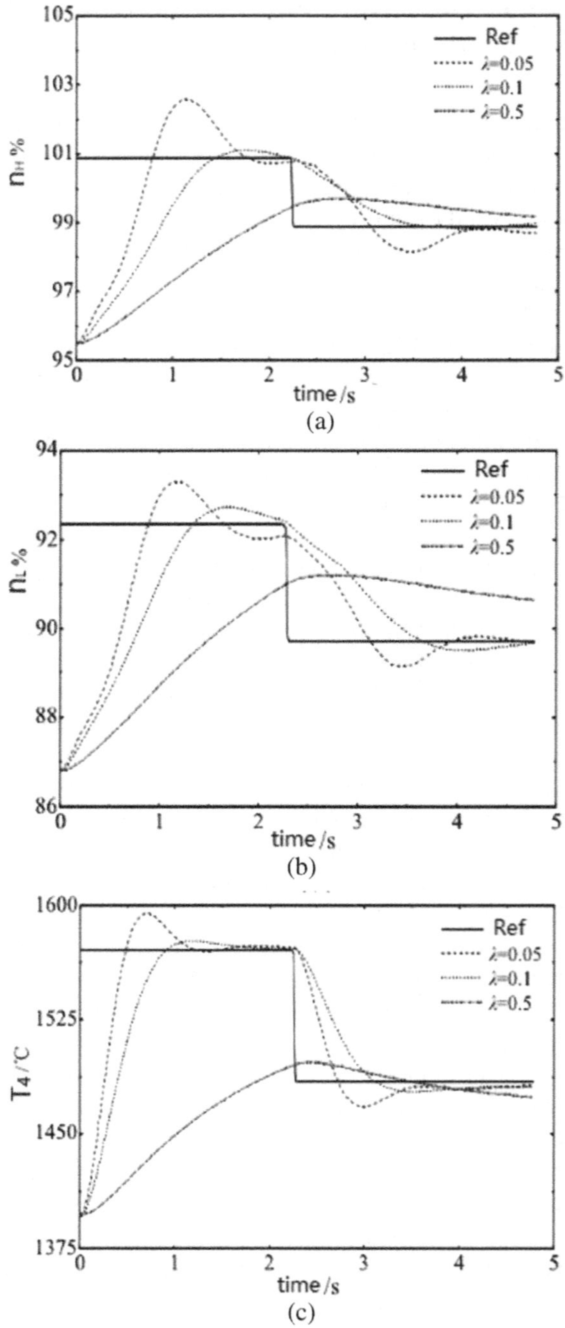


**Fig. 4.23** Aeroengine performance for different λ: **a** High-pressure rotational speed; **b** Low-pressure rotational speed; and **c** Turbine temperature

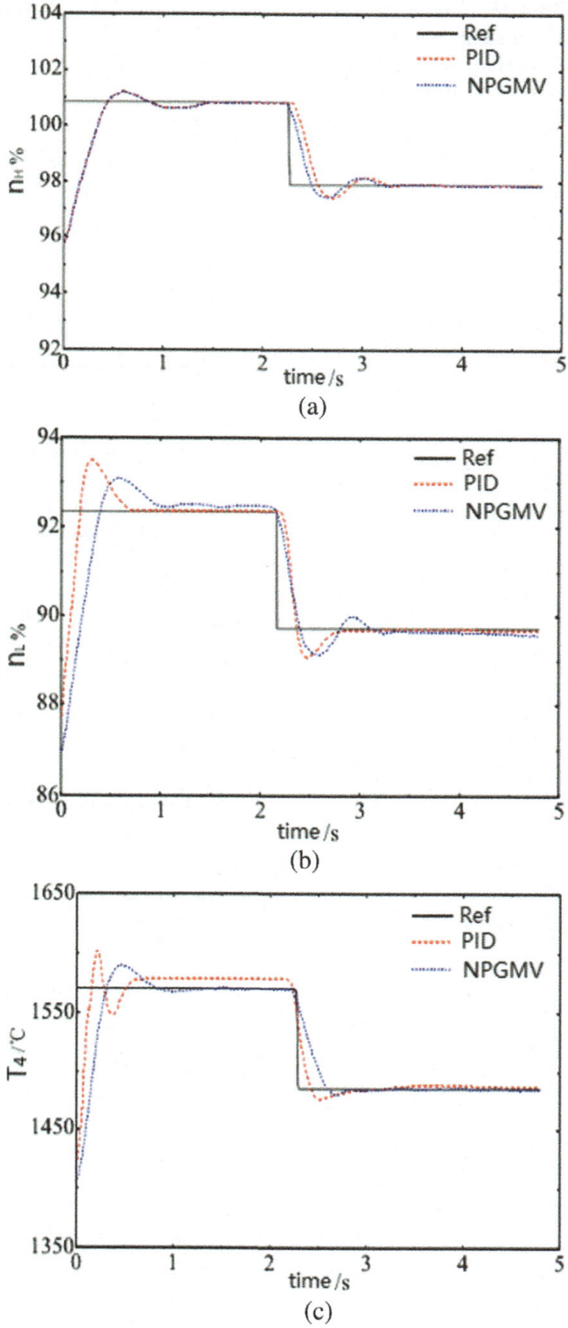


**Fig. 4.24** System performance comparison between PID and NPGMV: **a** High-pressure rotational speed; **b** Low-pressure rotational speed; and **c** Turbine temperature

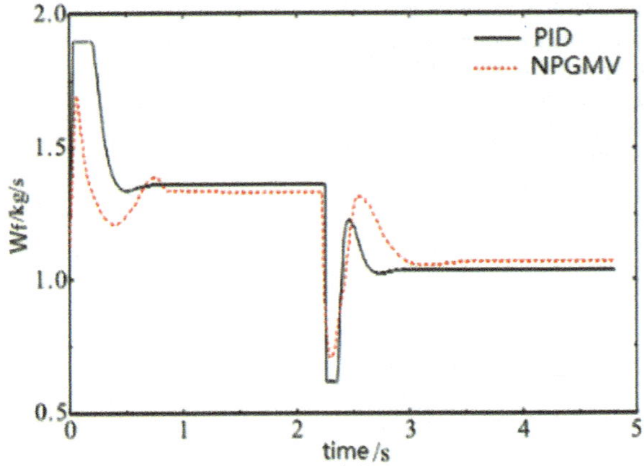

(a)

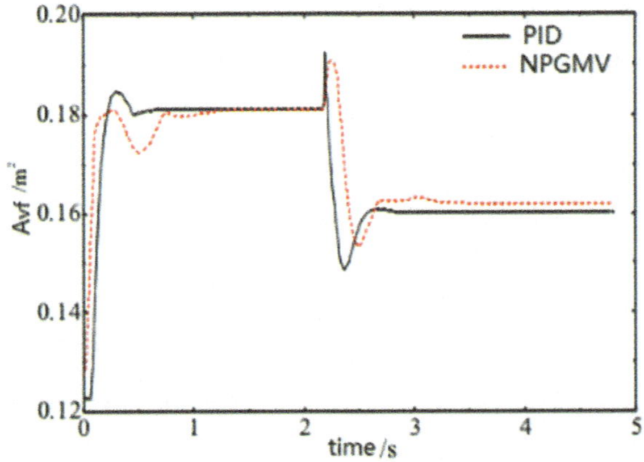

(b)

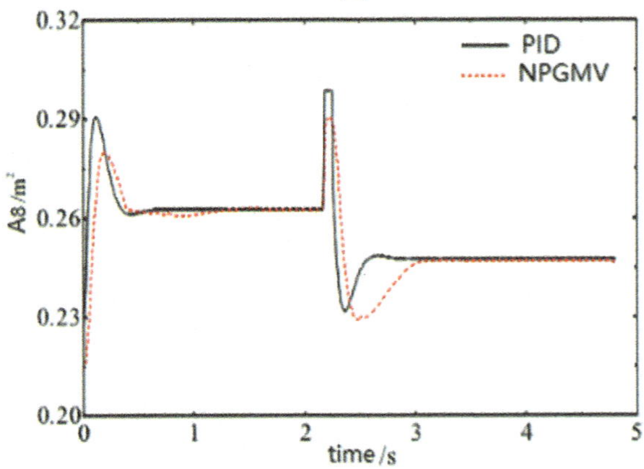

(c)

**Fig. 4.25** Control performance comparison between PID and NPGMV: **a** Fuel flow; **b** Area after the lower bypass; and **c**: Nozzle area control design. Indeed, fuel metering devices consist of springs in metering valves, and high-frequency, large magnitude actuation is not desired, since they can change the stiffness and force, henceforth reducing the dynamic quality of the actuation. In fact, this can further lead to hard fault in fuel metering devices, resulting in catastrophe for flight safety.

### *4.3.6 Summary for Nonlinear Predictive GMV Control of Aeroengines*

Nonlinear predictive generalized minimum variance control has been investigated for aeroengine transient control. Issues associated with performance index selection, optimal control law design and implementation, constraint handling, robustness and small control have also been explored and given careful exposition. Numerical examples have been extensively provided to for dissemination of this newly developed nonlinear transient control methodology. Results show that for the given example, this new proposal does possess superior property over the conventional PID control; meanwhile, unlike currently developed nonlinear model predictive control methodologies in the literature, the proposed NPGMV control does not rely on extensive computational resource and henceforth easy to design and detuning. Yet, it is quickly acknowledged that the weakness of NPGMV is its not capable of handling constraints in real-time application, in the sense that it is pre-calibrated and retains a fixed controller for application, thus not adaptive to un-prescribed situations. This issue will be further discussed in the next chapter for optimization-based transient control where real-time optimization is necessitated for constraints handling, and of course, with the sacrifice of computational resources.

## 4.4 Nonlinear Aeroengine Transient Control with Online Tuning

The above sections have presented both nonlinear generalized minimum variance-based control and its predictive counterpart for aeroengine transient control. One of the features of the approaches is that they do not require the explicit inversion of the nonlinear plant model to compute the optimal NGMV or NPGMV control law. More importantly, the NGMV and NPGMV control turn out to be applicable in a very wide range of nonlinear systems which even include those *systems modeled by look-up tables, neural-fuzzy network, or even old Fortran code.* This makes the NGMV and NPGMV control particularly suitable for industrial applications. It can even be related to a nonlinear version of Smith Predictor that is a tool for "every day use" in industry.

Meanwhile, for aeroengine set point control, it has been explained that set point control is essentially a "*disturbance rejection problem*" and the resulting design should provide fast regulation of the engine, e.g. be able to reject disturbances and robust against modelling uncertainties. For aeroengine transient control, however, disturbance should also be tolerated while retaining acceptable transient performance, even under possibly severe disturbances. In Sects. 4.1–4.3, disturbance uncertainty has been dealt as bounded disturbance and the optimal controllers are such designed to meet performance requirement even for the worst-case scenario [45–48]. This will inevitably introduce conservatism into the design and one way to avoid this is through disturbance observation—disturbance observer. This approach is powerful since the disturbance can be estimated so that "prognostic" action can be taken to reduce the conservatism. Therefore a further design freedom can be added in order to counteract the usually detrimental effect of disturbances in conjunction with feedback control [49]. It is worth mentioning that there are various forms of disturbance observers [50–52] which have found many practical applications (e.g. [53–55]). One of the disturbance observer approaches to robust control is active disturbance rejection control (ADRC) proposed by Han [56], consisting of tracking differentiation, extended state observer and a nonlinear PD. ADRC thus achieves robust control by compensating nonlinearity, model uncertainly and exogenous disturbances in real time.

There is yet another approach to robust control, namely reconfigurable control. Reconfigurable control in fact falls within the broad range of fault tolerant control systems, where robustness is achieved even in the event of malfunctions or failures in sensors, actuators or other components. As a consequence, reconfigurable control is often integrated with a fault detection process and performance is maintained through control reconfiguration, either pre-scheduled or online computation. Details can be referred to the excellent review articles [57, 58].

In this section, however, an online tuning method to robust control is proposed. On line tuning method is not newly invented, e.g. digital PID control used in many process industries are tuned/detuned in an online manner through monitoring the real time performance [59, 60]. The feature proposed here, however, is to tune the control performance through *predictive action*, e.g. the process is controlled by a nominal NGMV controller, and once deterioration is detected due to disturbances or other operational malfunctions, predictive action (NPGMV-type) is augmented to counteract the detrimental effects. Besides this, comparing with the existing robust control approaches, it has the following distinctive features:

(1) different from the disturbance observer-based approaches, the proposed method does not need real time estimation of disturbance and henceforth is computationally economic;
(2) different from reconfigurable control-based approaches, the proposed method does not require control reconfiguration mechanism and therefore preserves the hardware structure and topology of the system;
(3) different from all the existing approaches introduced above, the proposed method can target those complicated nonlinear systems with non-analytical

modules such as C/Fortran codes, lookup tables or even neural networks. This is so since it can be thought of as combination of NGMV and NPGMV control.

To fully comprehend the contribution, these features deserve further explanation by putting in the background of aeroengine control:

(1) disturbance observer usually needs extensive computation however in many systems computational resources are quite limited, e.g. in the aircraft engine electronic control unit (EEC), the sampling rate is only of 20 ms due to harsh working conditions and extreme reliability, and it is impossible to provide online estimation of disturbance. In fact, even for the limited computational resource, only 15% of it can be used for control purpose and the remaining has to be allocated for health monitoring functions;
(2) in reconfigurable control, system topology is usually changed particularly due to actuator failures. But in many engineering systems, particularly in aeroengine industry, it is often desirable to retain the control structure and it is hoped that control performance can be maintained only through additional control injection;
(3) in most of the above approaches without real time estimation, it is implicitly assumed that an analytical model can be obtained prior to control design. However, many processes in industries do not admit a mathematically analytical model as represented, e.g. difference/differential equations. For example, lookup tables are included in high-speed powertrain models; C or even Fortran codes are programmed to represent the models for helicopters or aeroengines.

In fact, this paper is motivated by the effort for handling those nonlinear systems with non-analytical subsystems through simple controls [61]. The method proposed here should represent a novel approach to robust control of multivariable nonlinear processes in general and not restricted in aeroengine industry. This section is structured as follows: Sect. 4.4.1 describes the formulation of the online tuning method, by introducing a predictive action for robustness by use of anti-causal operator; Sect. 4.4.2 analyses constraint handling and stability issues; Sect. 4.4.3 provides a numerical study for validation of the proposed method while Sect. 4.4.4 summarizes the section.

### 4.4.1 *Formulation of Online Tuning Optimal Control: Use of Anti-Causal Operator*

The nominal control comes from the NGMV controller that has been exposited in Sect. 4.1. That is, an optimal NGMV controller has been designed with the following representation reproduced here:

$$u_{opt}(t) = -\left\{F_{ck} - P_c C_0 T_0(k, z^{-1}) B_0 W_{1k}\right\}^{-1}$$
$$P_c\left[C_r^k x_r(t|t) - C_d^k x_d(t|t) - C_0 A_0^k \hat{x}_0(t|t)\right] \tag{4.66}$$

Or:

$$u_{opt}(t) = -F_{ck}^{-1} P_c \left[C_r^k x_r(t|t) - C_d^k x_d(t|t) - C_0 A_0^k \hat{x}_0(t|t) - C_0 T_0(k, z^{-1}) B_0 \left(W_{1k} u_{opt}\right)(t)\right] \quad (4.67)$$

Now consider the nonlinear system suffers disturbances such as environmental changes or parametric drifting, and this can cause detrimental effect on the performance. H∞ control or reconfigurable control should be considered for handling this situation. However, as explained in the introduction, an online tuning method is proposed here. The idea consists of an essential utilization of *anti-causal operator*. In specific, the design parameters $P_c$ and $F_{ck}$ are such chosen that $\left(F_{ck}^{-1} P_c\right)$ is anti-causal, and this implies that the optimal control signal as generated through Eq. (4.23) will includes two parts: a causal part $N_c(z^{-1})$ and an anti-causal part $N_{ac}(z^{-1})$. The optimal signal now reads:

$$u_{opt}(t) = \left[-N_c(z^{-1}) g(t)\right] + \left[-N_{ac}(z^{-1}) g(t)\right] \quad (4.68)$$

where $g(t) \equiv C_r^k x_r(t|t) - C_d^k x_d(t|t) - C_0 A_0^k \hat{x}_0(t|t) - C_0 T_0(k, z^{-1}) B_0 \left(W_{1k} u_{opt}\right)(t)$.

A schematic implementation of controller (4.68) is presented in Fig. 4.26b, which shows optimal signal generated with one-step ahead prediction. This is in comparison with the usual implementation by Eq. (4.67) in Fig. 4.26a.

It is noted from Fig. 4.26 that the implementation of the optimal control signal does not rely on the inversion of the nonlinear system. This substantially reduces the complexity and computational burden usually associated with nonlinear predictive control. But the much pertaining issue is that through such an implementation, predictive action can be easily augmented into the original nonlinear controller with the potential to improve performance.

To proceed, suppose a nominal controller has been designed as follows:

$$\begin{aligned} u(t) &= -N_c(z^{-1}) g(t) \\ &= -N_c(z^{-1}) \left\lfloor C_r^k x_r(t|t) - C_d^k x_d(t|t) - C_0 A_0^k \hat{x}_0(t|t) - C_0 T_0(k, z^{-1}) B_0 \left(W_{1k} u_{opt}\right)(t \right\rfloor \end{aligned} \quad (4.69)$$

Now it is known that this nominal design provides optimal control resulting in the nominal performance:

$$y(t) = d(t) + (Wu)(t) \quad (4.70)$$

The existence of, say, additive uncertainty $\Delta$ in plant can degrade the system performance:

$$y_p(t) = d(t) + [(W + \Delta)u](t) \quad (4.71)$$

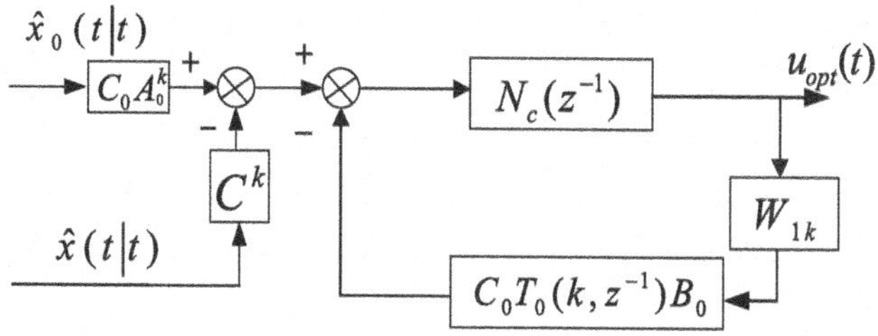


(a) optimal control signal as generated through (4.66)

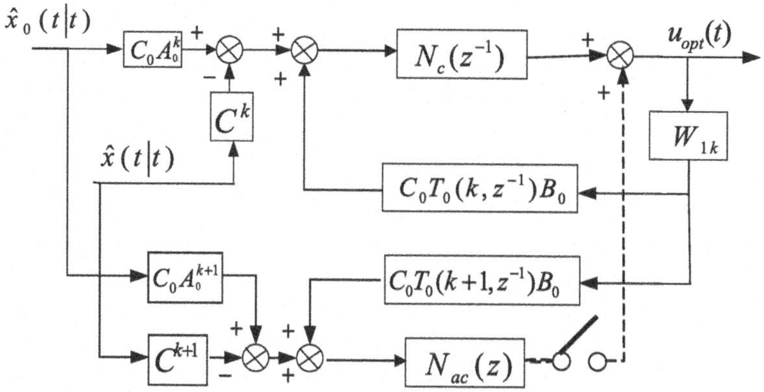


(b) optimal control signal with augmented predictive action as generated through (4.67)

**Fig. 4.26** Optimal signal generation with anti-causal parameters $\left(F_{ck}^{-1}P_c\right)$. One step ahead prediction is augmented into the usual implementation without anti-causal choice of $\left(F_{ck}^{-1}P_c\right)$

The idea of robust design is to utilize the predictive action to counteract the detrimental effects of system uncertainty, e.g. using:

$$u_p(t) = u(t) + \left[-N_{ac}(z^{-1})g(t)\right] \tag{4.72}$$

where $N_{ac}(z^{-1})$ is an anti-causal weighting matrix but not necessarily comes from the decomposition of $\left(F_{ck}^{-1}P_c\right)$. Referring to Eq. (4.68), this essentially means adding predictive action to the existing control $u(t)$. Now define $u_{pa}(t) = -N_{ac}(z^{-1})g(t)$ then Eq. (4.72) can be conveniently written as:

$$u_p(t) = u(t) + u_{pa}(t) \tag{4.73}$$

The performance now is:

$$y_p(t) = d(t) + \lfloor (W + \Delta)u_p \rfloor(t) \tag{4.74}$$

The condition that this performance should approximate the nominal performance is thus:

$$\left(\Delta u_p\right)(t) + \left(W u_{pa}\right)(t) = 0 \tag{4.75}$$

Equation (4.75) suggests a systematic tuning procedure for the "robustification" of the nominal controller using predictive control.

Referring to Fig. 4.27, a nominal controller $u(t)$ has been designed based on the model $W$. Before commissioning on the real plant, predictive action $u_{pa}(t)$ is introduced through the structure illustrated in Fig. 4.27. This generates a residual signal $\xi(t)$; tune $N_{ac}(z^{-1})$ to reduce this residual signal. The limiting case $E\{\xi(t)\} = 0$ indicates the satisfaction of Eq. (4.75). This consequently implies that the introduction of the predictive action $u_{pa}(t)$ has successfully offset the performance deterioration caused by the system uncertainty $\Delta$. As a result, the real system performance $y_p(t)$ will approximate the optimal nominal design $y(t)$.

**Remark 3** It is seen that the above procedure results in a two-step design methodology: in the first step, an optimal controller based on a nominal system model is designed; predictive action is then introduced to deal with the system uncertainties. The interesting thing about this design procedure is that the optimization of the cost function $J = E\{\phi^T(t)\phi(t)\} = E\{trace[\phi^T(t)\phi(t)]\}$ is maintained as can be deduced from Proposition 1, although this "*Augmentation Mechanism*" should be carefully considered to provide optimal performance.

**Remark 4** The proposed methodology can considerably simplify control system design since only the system model is actually required throughout the design process. It thus provides a simple on-line tuning of the optimal controller to handle modeling

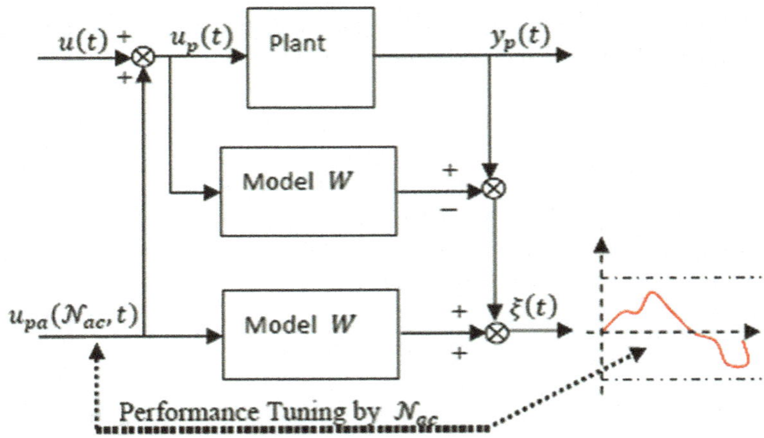


**Fig. 4.27** Robustified nominal controller: tuning the parameter $N_{ac}$ to reduce the residue signal $\xi(t)$. The resulting controller will counteract the detrimental effect of the system uncertainties

errors, while conventional analytical robust control design generally produces a conservative controller.

**Remark 5** Finally, it is worth pointing out that robustness analysis can also be conducted though the introduction of predictive control. To see this, consider the uncertainty $\Delta$ will generally be specified by the bound on its 2-gain, or extended *H∞ Norm*:

$$\|\Delta\|_{\infty} \leq \gamma \tag{4.76}$$

The necessary and sufficient condition for satisfying Eqs. (4.75) and (4.76) is:

$$\frac{\left\|(Wu_{pa})(t)\right\|_2}{\left\|u_p(t)\right\|_2} \leq \gamma \tag{4.77}$$

This serves to analyze the problem of tolerating the specified uncertainty bound.

### 4.4.2 Constraint Handling for Input Saturation and Stability Issues

One of the most important constraints for any control system design is the input constraint, e.g. actuators will eventually saturate. This can cause serious problem on system performance. The advantage of the proposed control is that it can handle the input saturation easily. This is so since the input saturation characteristic $Sat(\cdot)$ can be viewed as a nonlinear function of the input signal u(t)$u(t)$. This nonlinearity can then be included into the non-analytical module $W_{1k}$ (bear in mind this non-analytical module can be lookup tables, C/Fortran codes, neural networks etc.) as:

$$W_{1k}^S = [(W_{1k}Sat)u](t) \tag{4.78}$$

Then all the results derived above are valid with respect to the non-analytical module $W_{1k}^S$. In specific, the optimal control signals (4.66) and (4.67) hold for $W_{1k}^S$. Then the implementation scheme still applies and is clearly equivalent to simply applying the saturation characteristics to the optimal control signal $u(t)$. Therefore handling input saturation may only involve detuning the weighting matrices $P_c$ and $F_{ck}$, and no additional controller and computational complexities are required.

Unlike many control design approaches, e.g. those Lyapunov-based design methodologies, stability is not guaranteed for the proposed approach, therefore care must be exercised to provide closed-loop stability. However, practical design experiences show that it is in fact easy to stabilize the system once a PID controller is available. A design procedure is suggested to choose the design parameter $\left(F_{ck}^{-1}P_c\right)$ to be identical PID controller, then closed-loop stability can be achieved; detuning of the weighting matrices $P_c$ and $F_{ck}$ can then improve the performance substantially. This resembles the analysis for NGMV control.

**Remark 6** Although the stability issues are not mathematically addressed, the above design procedure can result in improved closed loop performance based on the PID controller, which is very convenient for *industrial applications beyond aeroengine industry*. But it is also noted that stability proofs can indeed be given based on small gain theorem and circle criterion. The authors have deduced many conditions for stability guarantee for the situations where $W_{ok}$ is either stable or unstable. The trouble with these sufficient conditions is that they might be difficult to check and verify, as is the case for many situations for nonlinear control system designs.

### 4.4.3 Nonlinear Transient Control with Online Tuning

Here a simple aeroengine model is considered to illustrate the basic concept of online tuning. As done in Sects. 4.2–4.4, the optimal nominal NGMV controller is first designed based on the nominal stabilizing PD controller $C_{PD}(z^{-1})$. That is, the dynamic weightings are initially chosen as $F_{ck} = -C_{PD}(z^{-1})$ and $P_C(z^{-1}) = I$. $F_{ck}$ is then fixed while $P_C(z^{-1})$ is freely chosen. A choice of $P_c(z^{-1}) = \frac{1.9}{1-0.2z^{-1}}$ results in the performance as shown in Fig. 4.28. Also shown is the performance of the nominal PD controller. From Fig. 4.28 it is seen that a simple choice of $P_C(z^{-1})$ improves the performance substantially. More importantly, experience shows that, although based on "trial-and-error," this step is very easy to achieve.

Now suppose the system model has un-modeled dynamics that can be represented by a dynamic disturbance, e.g. 10% of dynamic perturbation is injected. It is noted that this may be considered to be a parametric uncertainty and thus represents a very realistic situation. Due to the presence of the uncertainty, drastic performance deterioration is observed as in Fig. 4.29. The predictive action can then be introduced as explained in Sect. 4.4.1. In this study the simplest case of one-step prediction is considered. That is, only the future information $g(t+1)$ is utilized as in Eq. (4.68). In the notation of Eq. (4.68):

$$u_{opt}(t) = \left[-N_c(z^{-1})g(t)\right] + \left[-N_{ac}(z^{-1})g(t+1)\right] \tag{4.79}$$

The part $\left[-\mathrm{N_c}(z^{-1})\mathrm{g}(t+k)\right]\left[-N_c(z^{-1})g(t)\right]$ denotes the optimal control part. Predictive action is augmented by the choice of the following nonlinear control action:

$$N_{ac}(z^{-1})g(t+1) = -5000[g(t+1) + 0.2g(t)] \tag{4.80}$$

From Fig. 4.30, it is seen that inclusion of such a predictive control action counteracts effectively the system uncertainty and improves significantly the control performance.

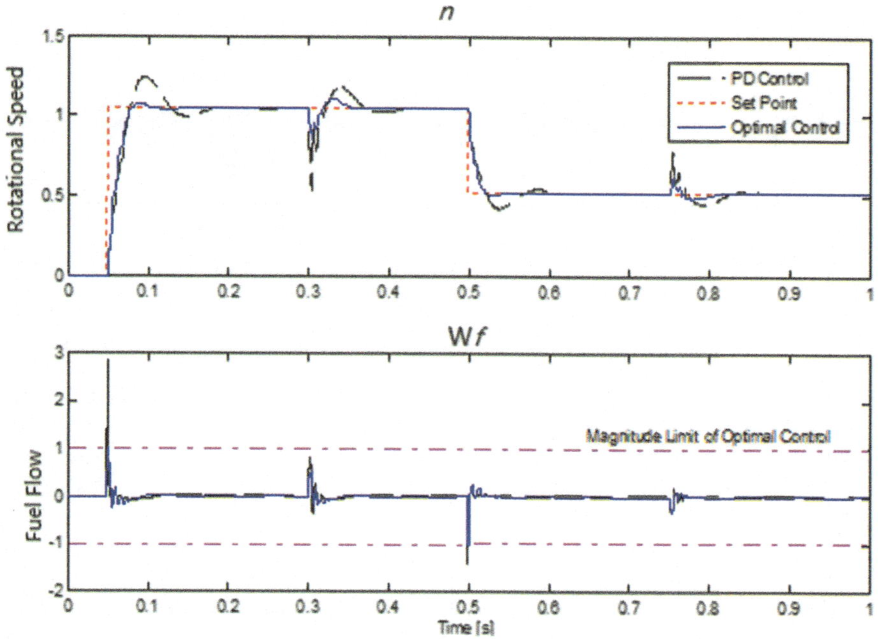


**Fig. 4.28** Rotational speed control with fuel flow as input: nominal control

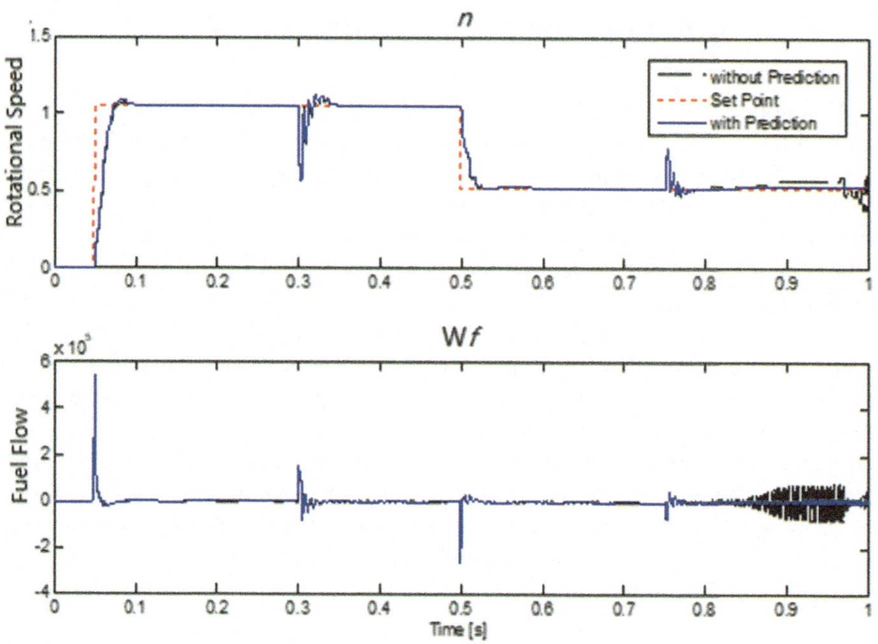


**Fig. 4.29** Rotational speed control with fuel flow as input: with prediction

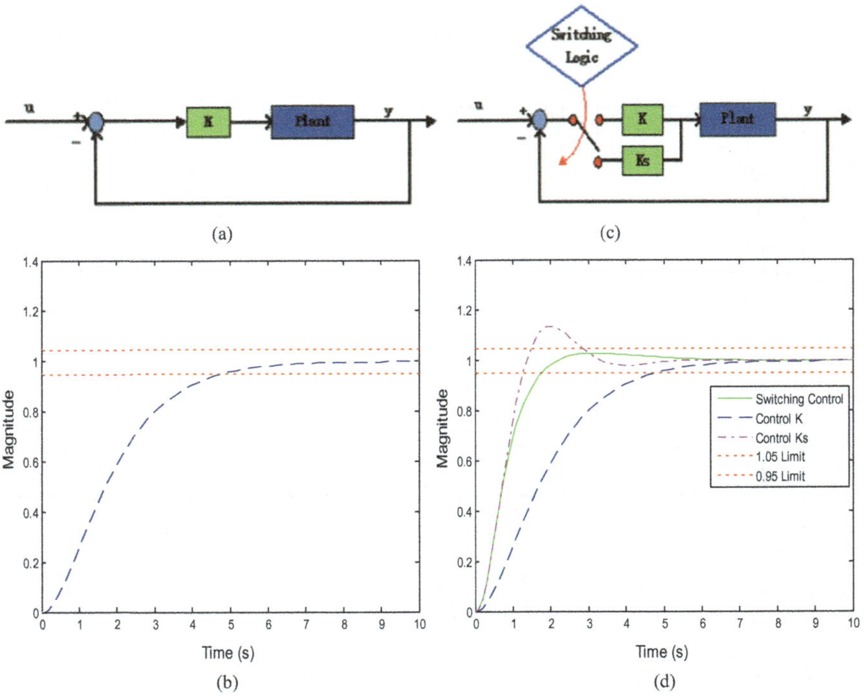


**Fig. 4.30** Concept of transient performance improvement through switching

Now notice from Fig. 4.29 that the magnitude of control signal of the PD controller is much larger than that of the proposed controller. This is deliberately designed to convince the reader that input saturation can be handled easily involving only a tuning of the weighting and no additional computational cost. As discussed in Sect. 4.4.2, this can be done by simply applying the saturation characteristics to the optimal control signal $u(t)$. This evidently illustrates the power of proposed control and its predictive extension.

Finally, it is stressed again that the predictive design can also be performed relatively easily and thus it is convenient for practical implementation. In conclusion, based on a nominal stabilizing *PID* controller, an optimal controller can be easily determined to improve system performance; the optimal control design can then be augmented to incorporate predictive control actions and this can result in the overall performance improvement even in the presence of dynamic uncertainties. Either the proposed optimal controller or its predictive extension can be easily detuned to handle input saturation without additional computational cost.

### 4.4.4 Summary for Nonlinear Aeroengine Transient Control with Online Tuning

This section introduces an online tuning method for fighting against uncertainty or disturbance in aeroengine transient control. The proposed method, although lack of a priori stability guarantee, is easy to implement and can be detuned to handle modelling uncertainties through a simple augmentation mechanism. Next step of research is to bring an adaptive feature for automatic augmentation of predictive action once performance deterioration is detected. This shall extend the current formulation to control reconfiguration in a fault tolerant control framework. As another line of research concerning practical applications, the proposed method echoes the performance recovery control for aeroengine control design, e.g. a primary control loop for fuel flow ratio, while by injection of mass flow or even water for additional propulsion force due to agile manoeuvres.

## 4.5 Switching-Based Aeroengine Transient Control

The above sections have completed a NGMV-based aeroengine transient control, and it is optimal in the sense that a prescribed performance index is to be minimized to obtain the optimal control. Almost as a diversion, before ending the story, an even simpler switching-based transient control method is introduced. This comes from the motivation and observation that all the control design methods aim to increase performance with some form of modification of plant property in one form or another. One question worth of asking is *when the parameters of controllers, sensors, and actuators are fixed, is it possible or how to improve the aeroengine transient performance*? This will bring the switching theory in a par.

Indeed, theory of switched systems has long been developed where many important results have been established [62–65]. The concept of switching control has been investigated in aeroengine multi-loop stability design [24, 66, 67], however, it will be shown that it can also be utilized for transient performance enhancement purpose. This will be beneficial for fixed system dynamics. The basic concept is shown in Fig. 4.30: given the aeroengine dynamics and controller $K$, the resulting closed loop control is shown in (a), with step response in (b). It is seen that the inertia causes sluggish response with a rising time of 5 s.

Thus, the basic idea is to first deploy a controller $Ks$ that can provide a fast transient response, see Fig. 4.30c; and the control authority is then handled back to the original controller $K$ near the desired steady state value. It should be noted, however, that the controller $K$ has a sluggish response, but the controller $Ks$ can lead to large overshoot if used alone. This is not acceptable due to the limit constraints (e.g. the high-pressure rotational speed should be within 0.5% overshoot at the design point). Refer to Fig. 4.30d, the controller $Ks$, when used alone, provides a fast transient response (2.64 s) but with a 20% overshoot; on the other hand, the switched controller has a 2.62 s settling time but without any overshoot. Thus, switching control improves transient performance. Consequently, the key issue is the design of an appropriate switching logic or switching condition.

### 4.5.1 Switching for Fast Transient Control: Design Methods

It is seen from above exposition that an appropriate switching logic should be sought. For aeroengine control systems, switching logic can be as simple as a threshold design, e.g. for the above step response, switching logic can be designed as output response reaching a prescribed threshold value. The design of threshold value can be obtained through trial-and error, but it can also be obtained using mathematical optimization techniques. For the case given above, suppose the regulation time $t_s$ is to be optimized. When only the controller $K$ is in authority, the system is critically damped with unity step response:

$$h(t) = 1 - e^{-\omega_n t}(1 + \omega_n t) \tag{4.81}$$

The resulting regulation time $\tilde{t}_{sK}$ is:

$$\tilde{t}_{sK} = \left\{ t \middle| e^{-\omega_n t}(1 + \omega_n t) = 0.05 \right\} \tag{4.82}$$

While only with $Ks$ in action, the system is under-damped, thus its unity step response is:

$$h(t) = 1 - \frac{1}{\sqrt{1-\xi^2}} e^{-\xi\omega_n t} \sin(\omega_n \sqrt{1-\xi^2} t + \arccos \xi) \tag{4.83}$$

The corresponding regulation time $\tilde{t}_{sKs}$ is:

$$\tilde{t}_{sKs} = \frac{\pi - \arccos \xi}{\omega_n \sqrt{1-\xi^2}} \tag{4.84}$$

Now consider the switched system whose response consists of two stages: stage I takes the output $y \in \left[0,\ 1-e\right]$ with response time $t_{sKs}$ and stage II after switching to $K$ to steady state with response time $t_{sK}$ for output $y \in \left[1-e,\ 1\right]$. The situation is illustrated in Fig. 4.31.

Therefore, under $Ks$, the time $t_{sKs}$ taking the output $y$ from zero to (1-$e$) is:

$$t_{sKs} = \left\{ t \middle| 1 - \frac{1}{\sqrt{1-\xi^2}} e^{-\xi\omega_n t} \sin(\omega_n \sqrt{1-\xi^2} t + \arccos \xi) = 1 - e \right\} \tag{4.85}$$

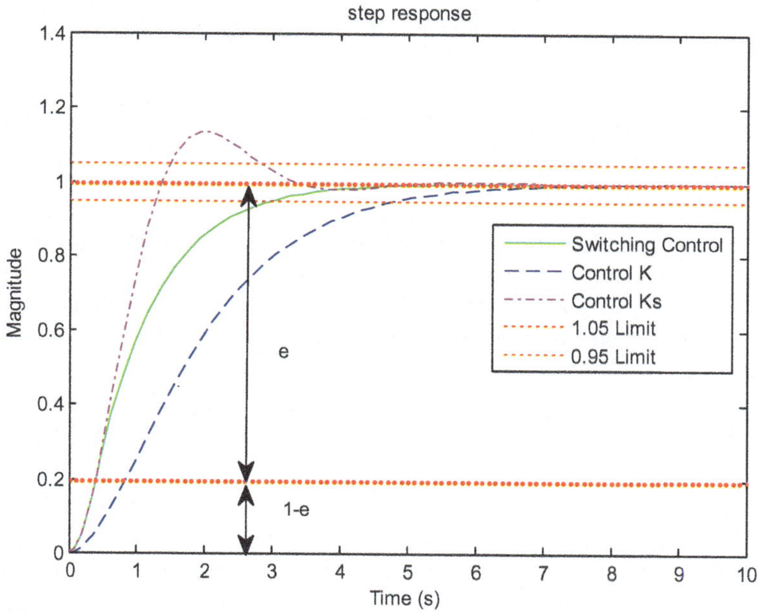


**Fig. 4.31** Key issue for switched system is the design of threshold value

After $Ks$ acting $t_{sKs}$ duration, controller $Ks$ is switched to controller $K$, the system is reset into the new initial condition to be obtained by the solution to the following equations:

$$
\begin{aligned}
y_e &= 1 - e \\
\dot{y}_e &= \frac{1}{\sqrt{1-\xi^2}} \\
&\Big[\xi\omega_n e^{-\xi\omega_n t_{sKs}} \sin(\omega_n\sqrt{1-\xi^2}t_{sKs} + \arccos\xi) \\
&-\omega_n\sqrt{1-\xi^2}e^{-\xi\omega_n t_{sKs}} \cos(\omega_n\sqrt{1-\xi^2}t_{sKs} + \arccos\xi)\Big]
\end{aligned}
\tag{4.86}
$$

Without loss of generality, define the dynamical equation corresponding to the closed-loop transfer function $G_{closed} = \frac{GK}{1+GK}$ under controller $K$ as:

$$
\begin{aligned}
\dot{x}(t) &= Ax(t) + Bu(t) \\
y(t) &= Cx(t)
\end{aligned}
\tag{4.87}
$$

It is relatively easy to calculate (4.87) with (4.86) as initial condition, the time duration $t_{sK}$ from $y = 1 - e$ to $y = 1$:

$$t_{sK} = \left\{ t \middle| Ce^{At}\left[\begin{bmatrix} y_e \\ \dot{y}_e \end{bmatrix} + \int_0^t e^{-A\tau} Bu(\tau)d\tau\right] = 1 \right\} \tag{4.88}$$

The total regulation time for the switched control systems is thus:

$$t_s = t_{sKs} + t_{sK} \tag{4.89}$$

And the switched control design problem can now be transformed into the following optimization problem:

***Optimization problem for regulation time $t_s$ with switched control: find a threshold value e, such that the regulation time $t_s$ is optimized.***

That is:

$$\min_{e\in(0,\,1)} \quad t_s$$

$$s.t. \quad t_s = t_{sKs} + t_{sK};$$

$$t_{sKs} = \left\{ t \middle| 1 - \frac{1}{\sqrt{1-\xi^2}} e^{-\xi\omega_n t} \sin(\omega_n\sqrt{1-\xi^2}t + \arccos\xi) = 1 - e \right\}$$

$$t_{sK} = \left\{ t \middle| Ce^{At}\left[\begin{bmatrix} y_e \\ \dot{y}_e \end{bmatrix} + \int_0^t e^{-A\tau} Bu(\tau)d\tau\right] = 1 \right\}$$

where

$$y_e = 1 - e;$$

$$\dot{y}_e = \frac{1}{\sqrt{1-\xi^2}}$$

$$\left[\xi\omega_n e^{-\xi\omega_n t_{sKs}} \sin(\omega_n\sqrt{1-\xi^2}t_{sKs} + \arccos\xi) - \omega_n\sqrt{1-\xi^2}\right.$$

$$\left. e^{-\xi\omega_n t_{sKs}} \cos(\omega_n\sqrt{1-\xi^2}t_{sKs} + \arccos\xi)\right]$$

However, it must be pointed out that it is not a easy task to solve the above optimization problem: observe the expression of $t_s$, it is seen that $t_s$ is not an explicit function of $e$, thus frustrating the conventional methods for solving optimization problems. For this situation, trial-and-error can still be a feasible method, e.g. choose a set of threshold values $e$, and different threshold produces different response, the optimal threshold $e_{opt}$ is determined with the optimal response. In fact, due to the change of operation conditions and environmental uncertainty, it is not practically appealing to optimize $t_s$ based only on nominal $G$. The trial-and-error approach is easy to implement, and transient response can be relatively easily improved through switched control.

### 4.5.2 Switching for Fast Transient Control: Numerical Study

Suppose for certain operation condition, the engine model with actuator is represented by $G = \frac{1}{s}\frac{1}{s+2}$ (hydraulic actuators can be approximated by an integrator element), and the controller is a simple proportional element $K = 1$, with switched controller $Ks = 3.5$. When the threshold values are chosen to be 0.8, 0.6, 0.4, and 0.2, the corresponding switched system response is illustrated in Fig. 4.32.

Figure 4.32 clearly illustrates the switched system response for different threshold values. The following observations can be obtained:

(1) The rising time $t_r$ of switched system is in between the responses of the two individual controllers;

(2) The settling time $t_s$ of switched system can be better than those of individual controllers, e.g. for threshold value $e = 0.6$, the switched system responses with a settling time only 1.5 s, while two individual controllers produce settling time

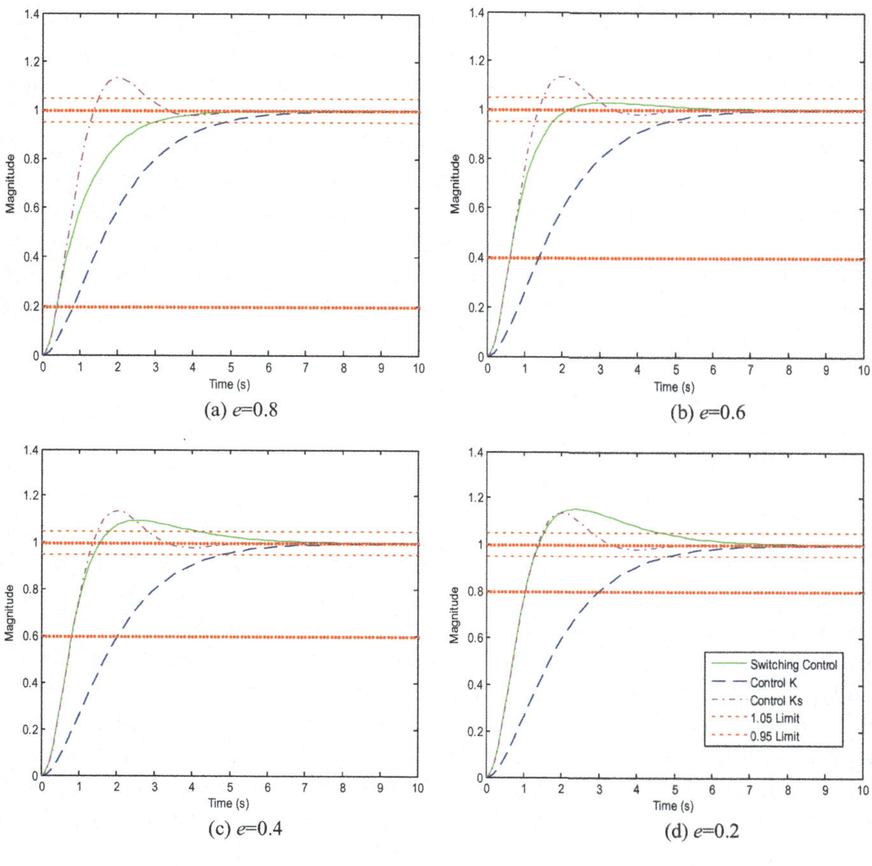


(a) $e$=0.8 (b) $e$=0.6 (c) $e$=0.4 (d) $e$=0.2

**Fig. 4.32** Switched system response for different threshold values of 4.8 s and 2.7 s, respectively. However, an inappropriate threshold value can cause an unfortunate situation, e.g. the settling time reaches to 5 s for $e = 0.2$.

(3) Overshoot $\sigma\%$ resembles the same characteristics as settling time $t_s$.

### *4.5.3 Switching for Fast Transient Control: Numerical Study*

A simple method for performance improvement using switching control has been introduced. In aeroengine control systems, actuator and sensor inertia can have significant impact upon control performance. Consequently, it is extremely important to develop high-precision and high-response sensing and actuation devices for performance enhancement [68–74]. However, for situations where the new technology readiness level is not ready, amelioration of mature technology is always worthwhile, particularly for aviation industry where safety and reliability are primary requirements [75–77]. The proposed switching control technique can improve transient performance even when the parameters of controllers, sensors, and actuators are fixed. Although it also brings the problem of increasing switching logic, it does possess the feature of ease of design and detuning. Consequently, the proposed method provides important insight into control performance enhancement with low cost and low complexity solutions.

Finally, it is worth pointing out that this problem comes from the authors' practical experience in engine control design, where the components inertia hurdles performance and causes significant damage to turbine blades. Quite a several schemes are devised, e.g. blades coating, spray nozzle re-design, high-pressure compressor air pre-bleed etc., but it is finally realized that the time and pressure delay associated with components inertia are best handled through FADEC laws, and henceforth the current paper for a solution.

## 4.6 Summary

The requirement for aeroengine transient control is to have a safe and fast transient response when manoeuvring. To achieve this, survey has been conducted in this chapter, starting from "standard practice" in industry as schedule-based transient control and acceleration-based control; moving to the introduction to linear parametric varying control and other advanced control techniques. However, considering the key feature of aeroengine transient process as "no-analytical," this chapter has initiated a systematic attempt with a nonlinear generalized minimum variance-based aeroengine transient control. Specifically, basic NGMV control and its predictive counterpart are developed; integration of both to achieve robustness against disturbance has been introduced. Finally, to conquer the issue of sensor/actuator inertia, as well as addressing the fixed controller problem, a switching-based aeroengine transient control design method has also been presented. The last method, although rule-of-thumb, does provide a simple, economic, easy-to-detuning solution to fimprove the transient performance of aeroengines.

## References


1. Jaw, L.C., and J.D. Mattingly. 2009. *Aircraft engine controls: Design, system analysis and health monitoring*, AIAA, Inc., Virginia.
2. MacIsaak, B.D., and H.I.H. Saravanamutto. 1974. A comparison of analog, digital and hybrid computing techniques for simulation of gas turbine performance. ASME Paper No. 74-GT-127.
3. Walsh, P.P., and P. Fletcher .2002. Gas turbine performance, blackwell science, USA, 1998. In *Two-spool, turbo-fan engine: (a) steady-state and dynamic mathematical models*, ed. Lichtsinder, M., and Y. Levy, 20–37. Technion, Israel, TAE No. 903.
4. Khalid, S.J., and R.N. Hearne. 1980. Enhancing dynamic model fidelity for improved prediction of turbofan engine transient performance. AIAA 80–1083.
5. Lichtsinder, M., and Y. Levy. 2006. Jet engine model for control and real-time simulations. *ASME Journal of Engineering for Gas Turbines and Power* 128:745–753.
6. Martin, S., I. Wallace, and Bates, D.G. 2008. Development and validation of a civil aircraft engine simulation model for advanced controller design. *ASME Journal of Engineering for Gas Turbines and Power* 130(5):051601:1–15.
7. Kreiner, A., and K. Lietzau. 2001. The use of onboard real-time models for jet engine control, MTU Aero Engines, Technical Report.
8. Isidori, A. 1995. *Nonlinear control systems*, 3rd ed. Berlin: Springer.
9. Van der Shaft, A.J. 1996. *L2 Gain and passivity techniques in nonlinear control*. Heidelberg: Springer.
10. Byrnes, C.I. 2000. *Toward a nonequilibrium theory for nonlinear control systems*. Lecture Notes in Control and Information Sciences. Berlin: Springer.
11. Sepulchre, R., M. Jankovic, and P.V. Kokotovic. 1997. *Constructive nonlinear control*. New York: Springer.
12. Block, A.M., J. Baillieul, P. Crouch and J. Marsden. 2003. In *Nonholonomic mechanics and control*, Springer.
13. Mayne, D.Q., J.B. Rawlings, C.V. Rao, and P.O.M. Scokaet. 2000. Constrained model predictive control: Stability and optimality. *Automatica* 36:787–814.
14. Apkarian, P., P. Gahinet, and G. Becker. 1995. Self-scheduled H∞ control of linear parameter-varying systems: A design example. *Automatica* 31:1251–1261.
15. Wu, F., X. Yang, A. Packard, and G. Becker. 1996. Induced L2-norm control for LPV systems with bounded parameter variation rates. *International Journal of Robust and Nonlinear Control* 6:983–998.
16. Stilwell, D., and W. Rugh. 2000. Stability preserving interpolation methods for the synthesis of gain scheduled controllers. *Automatica* 36:665–671.
17. Prakash, R., S.V. Rao, and C.A. Skarvan. 1990. Life enhancement of a gas turbine engine by temperature control. *International Journal of Turbo and Jet Engines* 7:187–206.
18. Wang, J., Z. Ye, and Z. Hu. 2012. Nonlinear control of aircraft engines using a generalized Gronwall-Bellman lemma approach. *ASME Journal of Engineering for Gas Turbines and Power* 134(9):094502.
19. Balas, G. 2002. Linear parameter-varying control and its application to a turbofan engine. *International Journal of Robust and Nonlinear Control* 12(9):763–793.
20. Bruzelius, F. 2004. In *Linear parameter-varying systems-an approach to gain scheduling*. PhD Thesis, Chalmers University, Göteborg, Sweden.

21. Henrion, D., L. Reberga, J. Bernussou, and F. Vary. 2004. Linearization and identification of aircraft turbofan engine models. In *Proceedings of IFAC Symposium on Automatic Control in Aerospace*, St. Petersburg, Russia.
22. Reberga, L., D. Henrion, J. Bernussou, and F. Vary. 2005. LPV modeling of a turbofan engine. In *Proceedings of IFAC World Congress on Automatic Control in Aerospace*, Prague, Czech Republic.
23. Vary, F., and L. Reberga. 2005. Programming and computing tools for jet engine control design. In *Proceedings of IFAC World Congress on Automatic Control*, Prague, Czech Republic.
24. Gilbert, W., D. Henrion, J. Bernussou, and D. Boyer. 2010. Polynomial LPV synthesis applied to turbofan engines. *Control Engineering Practice* 18:1077–1083.
25. Yao, Y., and J. Sun. 2008. Aeroengine direct thrust control based on neural network inverse control. *Journal of Propulsion Technology* 29(2):249–252.
26. Qi, X., and D. Fan. 2005. Application of improved FSQP algorithm to turbofan engine nonlinear multivariable control. *Journal of Propulsion Technology* 26(1):58–61.
27. Brunell, B.J., R.R. Bitmead, and A.J. Connolly. 2002. Nonlinear model predictive control of an aircraft gas turbine engine. In *Proceedings of 41st IEEE Conference on Decision and Control*, Las Vegas, USA.
28. Cutler, C.R., and B.L. Ramaker. 1979. Dynamic matrix control— A computer control algorithm. In *AIChE 86th National Meeting*, Houston, TX.
29. Cutler, C.R., and B.L. Ramaker. 1980. Dynamic matrix control—A computer control algorithm. In *Joint Automatic Control Conference*, San Francisco, CA.
30. Clarke, D.W., C. Mohtadi, and P.S. Tuffs. 1987. Generalized predictive control—Part I: The basic algorithm. *Automatica* 23:137–148.
31. Clarke, D.W., C. Mohtadi, and P.S. Tuffs. 1987. Generalized predictive control—Part II: Extensions and interpretations. *Automatica* 23:149–160.
32. Qin, S.J., and T.A. Badgwell. 1996. An overview of industrial predictive control technology. In *Proceedings of the 5th International Conference on Chemical Process Control (CPC-V)*, vol. 93, ed. Kantor, J.C., C., Garcia, E., and B. Carnahan, 232–256. Tahoe City, CA, AIChE Symposium Series No. 316.
33. Morari, M., and J.H. Lee. 1999. Model predictive control: Past, present and future. *Computers and Chemical Engineering* 23:667–682.
34. Meadows, E.S., and J.B. Rawlings. 1993. *Receding horizon control with an infinite horizon*, 2926–2930. San Francisco: In Proc. ACC.
35. Chen, H., and F. Allgöwer. 1998. A quasi-infinite horizon non-linear model predictive control scheme with guaranteed stability. *Automatica* 34:1205–1217.
36. De Oliveira, S.L., and M. Morari. 2000. Contractive model predictive control for constrainted nonlinear systems. *IEEE Transactions on Automatic Control* 45(6):1053–1071.
37. Mayne, D.Q., J.B. Rawlings, C.V. Rao, and P.O.M. Scokaert. 2000. Constrained model predictive control: Stability and optimality. *Automatica* 26(6):789–814.
38. De Nicolao, G., L. Magni and R. Scattolini. 2000. Stability and robustness of nonlinear receding horizon control. In *Nonlinear Predictive Control*, ed. Allgöwer, F., and A. Zheng, 3–23. Birkhäuser.
39. Rawlinigs, J.B. 2000. Tutorial overview of model predictive control. *IEEE Control System Magazine* 20(3):38–52.
40. Qin, S.J., and T.A. Badgwell. 2003. A survey of industrial model predictive control technology. *Control Engineering Practice* 11:733–764.
41. Grimble, M.J. 2008. Simple polynomial approach to nonlinear control. In *2008 CDC Symposium*, Cancun, Mexico to honor Bill Wolovich, Dec, 7.
42. Grimble, M.J. 2005. Nonlinear generalized minimum variance feedback, feedforward and tracking control. *Automatica* 41:957–969.
43. Grimble, M.J. 2007. In *Inviated plenary: Why industry needs improved and practical nonlinear control: A challenge for today*, IEEE Mediterranean Conference on Control and Automation, Athens, Greece, 27–29 June.

44. M.J. Grimble. 2008. Nonlinear predictive GMV control. In *2008 American Control Conference*, Westin Seattle Hotel, Seattle, Washington, USA, 11–13 June.
45. Zhou, K., J.C. Doyle, and K. Glover. 1996. In *Robust and Optimal Control*. Prentice Hall.
46. Kristić, M., J. Sun, and P.V. Kokotović. 1996. Robust control of nonlinear systems with input unmodeled dynamics. *IEEE Transactions on Automatic Control* 41:913–920.
47. Ahmadi, M., H. Mojallali, and R. Wisniewski. 2012. Robust $H\infty$ control of uncertain switched systems defined on polyhedral sets with Filippov solutions. *ISA Transactions* 51(6):722–731.
48. de Best, J., B. Bukkems, M. van de Molengraft, W. Heemels, and M. Steinbuch. 2008. Robust control of piecewise linear systems: A case study in sheet flow control. *Control Engineering Practice* 16:991–1003.
49. Franklin, G.F., J.D. Powell, and A. Emami-Naeini. 2002. In *Feedback control of dynamical systems*. 4th ed. Addison-Wesley Longman Publishing Co. Inc.
50. Radke, A., and Z. Gao. 2006. A survey of state and disturbance observers for practitioners. In *Proceedings of the 2006 American Control Conference*, pp. 5183–5188. 1–42440209–3/06, USA, Minnesota, June 2006.
51. Li, S., J. Yang, W. Chen and X. Chen. 2014. *Disturbance observer based control: Methods and applications*. CRC Press.
52. Choi, Y., K. Yang, W.K. Chung, et al. 2003. On the robustness and performance of disturbance observers for second-order systems. *IEEE Transactions on Automatic Control* 48(2):315–320.
53. Choi, H.T., B.K. Kim, and I.H. Suh et al. 2000. Design of robust high-speed motion controller for a plant with actuator saturation. *Transactions of the ASME, Journal of Dynamic Systems, Measurement and Control* 122(3):535–541.
54. Lee, K.B., and F. Blaabjerg. 2007. Robust and stable disturbance observer of servo system for low-speed operation. *IEEE Transactions on Industry Applications* 43(3):627–635.
55. Yang, J., A. Zolotas, W. Chen, K. Michail, and S. Li. 2011. Robust control of nonlinear MAGLEV suspension system with mismatched uncertainties via DOBC approach. *ISA Transactions* 50(3):389–396.
56. Han, J. 2009. From PID to active disturbance rejection control. *IEEE Transactions on Industrial Electronics* 56(3):900–906.
57. Zhang, Y., and J. Jiang. 2008. Bibliographical review on reconfigurable fault-tolerant control systems. *Annual Review in Control* 32:229–252.
58. Jiang, J., and X. Yu. 2012. Fault-tolerant control systems: A comparative study between active and passive approaches. *Annual Reviews in Control* 36:60–72.
59. Dey, C., and R. Mudi. 2009. An improved auto-tuning scheme for PID controllers. *ISA Transactions* 48(4):396–409.
60. Das, S., S. Saha, S.P. Das, and A. Gupta. 2011. On the selection of tuning methodology of FOPID controllers for the control of higher order processes. *ISA Transactions* 50(3):376–388.
61. Wang, J., Z. Hu, and Z. Ye. 2014. Static feedback stabilization of nonlinear systems with single sensor and single actuator. *ISA Transactions* 53:50–55.
62. Lin, H., and P. Antsaklis. 2009. Stability and stabilizability of switched linear systems: A survey of recent results. *IEEE Transactions on Automatic Control* 54(2):308–322.
63. Boscain, U. 2008. A review on stability of switched systems for arbitrary switching. In *Series on Advances in Mathematics for Applied Sciences* 76: Geometric Control and Nonsmooth Analysis, pp. 100–119.
64. Branicky, M.S. 1998. Multiple Lyapunov functions and other analysis tools for switched and hybrid systems. *IEEE Transactions on Automatic Control* 43(4):475–482.
65. Cheng, D.Z., L. Guo, Y. Lin, and Y. Wang. 2005. Stabilization of switched linear systems. *IEEE Transactions on Automatic Control* 50(8):1224–1228.
66. Liu, X. 2011. Switching control system design for aero-engine based on mixed sensitivity. *Journal of Aerospace Power* 26(6):1389–1395.
67. Yu, D., X. Liu, and Y. Sui. 2007. Design of aero engine multi-loop controllers with switching characteristics. *Journal of Aerospace Power* 22(8):1378–1383.
68. Jaw, J. C., and S. Garg. 2005. Propulsion control technology development in the US. *NASA/TM-2005–213978*.

69. Ballal, D., and J. Zelina. 2004. Progress in aeroengine technology. *Journal of Aircraft* 41(1):43–50.
70. Zhang, S. 2004. A review of aeroengine control system. *Journal of Aerospace Power* 19(3):375–382.
71. Paemel, M. 1989. Interface circuit for capacitive accelerometer. *Sensor and Actuators* 17:629–637.
72. Zhu, Z., and J. Ding. 2008. Research progress of dynamic characteristics of sensor. *Mechanics and Electronics* 6:75–78.
73. Li, J., D. Fan, C. Ji, and Z. Bu. 2012. Dynamic response capability improvement method for aero-engine temperature sensor. *Journal of Aerospace Power* 27(3):707–714.
74. Qian, K., S. Xie, and X. He. 2005. Aero-engine temperature sensor based on surging sign data fusion. *Journal of Propulsion Technology* 26(4):339–343.
75. Sun, J. 2001. Perspectives on aeroengine control in 21st century. *Journal of Aerospace Power* 16(2):97–102.
76. Yan, C. 2005. Some fundamental experiences of design and development of the "Kunlun" aero-engine. *Journal of Aerospace Power* 20(2):169–176.
77. Tulpule, B., A. Behbahani, and R. Millar. 2007. Vision for next generation modular adaptive generic integrated controls for military/commercial turbine engines. In *43rd AIAA/ASME/SAE/ASEE Joint Propulsion Conference and Exhibit*, Cincinnati, OH.

# Chapter 5
# Optimization-Based Aeroengine Nonlinear Control Integration

Aeroengine transient control has been analysed and one of the most important requirements for transient control is the safe and quick response to power request. Theoretically, the optimal operation is to follow the boundary of surge line (with surge margin) for acceleration while decreasing along burner blow-out limit for deceleration. However, in practice, surge and blow-out boundaries are extremely difficult to determine, particularly due to performance drift over flight envelope as well as engine usage cycle. Indeed, among the three basic functions (set point control, transient control and limit protection control) of engine control systems, transient control takes around three quarters of total development effort [1].

To handle this difficult transient control problem, as explained in the last chapter, the standard practice is through gain scheduling. That is, the flight envelope is divided into multiple regions where separate controllers are designed through conventional methods (usually PI controls); rapid engine acceleration and deceleration operation is achieved through scheduling variables (e.g. inlet sectional parameters and shaft rotational speeds) over full flight envelope. From implementation perspective, this gain scheduling method often results in a lookup table where the control variables such as fuel flow, guide vane angles etc. are operated in an open-loop manner. Varieties do exist for gain scheduling such as schedule-based control, acceleration-based control or $N$-dot control, but they are essentially obtained through experimentations and trials-and-errors. In fact, there are indeed problems associated with gain scheduling control: although dynamics and steady-state performance can be guaranteed at each operating point, transient performance may deteriorate (even lead to instability) during switching phase [2]. To handle this situation, anti-windup design is usually employed to accomplish "bumpless transfer" [3–5]. But it is noted that this bumpless transfer also implies parametric detuning. This is often not practical due to the reason that the parameters are obtained through extensive and costly experimentations. It is not desirable to be revised during commissioning phase. This is the reason behind the introduction of linear parameter varying (LPV) to engine transient design.

 
J. Wang et al., *Model-based Nonlinear Control of Aeroengines*,
https://doi.org/10.1007/978-981-16-4453-5_5

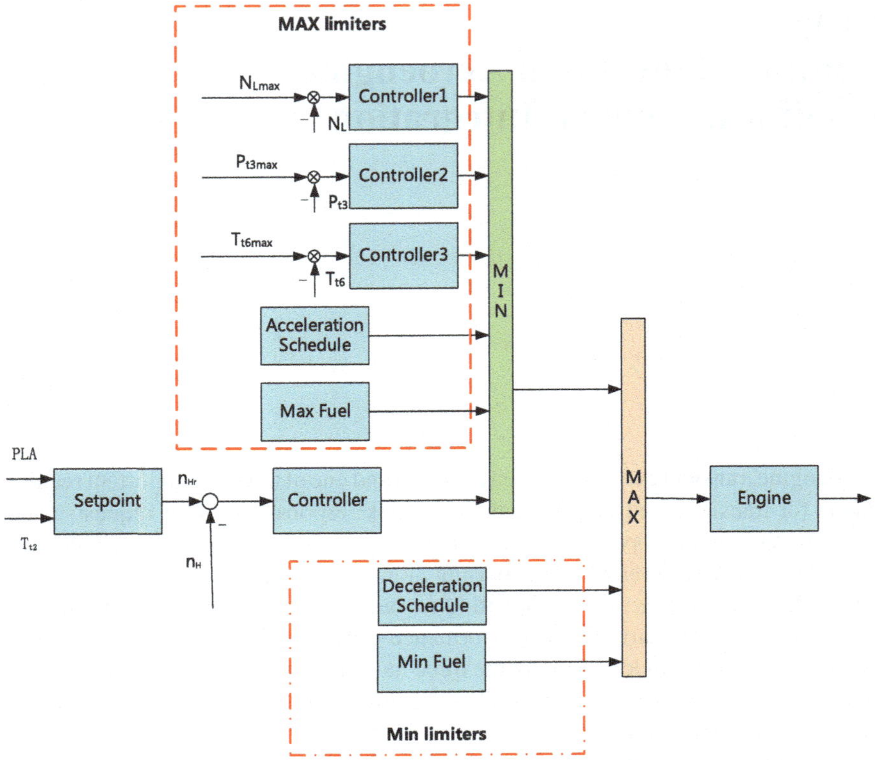


**Fig. 5.1** Control integration with limit protection control logic—limiters

Linear parameter varying control uses continuous scheduling of variable parameters to "prevent" both switching and detuning. The corresponding control theories have been well-established [6–11], while their application to engine control design has also been explored [12–17]. However, the linear parameter varying control does not provide guarantee for over-limits protection, e.g. engine should be protected from running beyond operating limits such as protection from undesirable operating conditions like stall, surge; not exceeding the maximum low-pressure and high-pressure shafts physical speed limit (afterburner nozzle area opens too fast or flame out); not exceeding the maximum compressor exist static pressure and temperature limits etc. While conventional approach to limit protection control is to use limiters as shown in Fig. 5.1, one advanced way to resolve this problem is to invoke those optimization-based methods. Here it is reminded that the development in the last chapter for NGMV and NPGMV-based controls are also not tailoring for integration of transient control with limit protection control.

Optimization-based methods include a variety of design approaches, with model predictive control [18, 19] and multi-objective and evolutionary algorithms [20, 21] as typical examples. These approaches have been demonstrated to have improved performances over conventional PID controls, with the special feature of ensuring not exceeding any operational limits. However, their power is usually exercised for control over set-point conditions (e.g. performance seeking control) or over a limited flight envelope. While transient control often refers to large envelope operations, new methodology has to be devised to account for significant change of engine dynamics over the envelope, as encountered in parametric varying control problems.

One approach to resolve this difficult large envelope control problem is to utilize a "piecewise optimization" technique. The fundamental idea is to partition the transition process into a relatively large number of "pieces" and within each piece, the engine dynamics is approximated by a linear process; the optimal control solution is then obtained based on the liberalized dynamics. The final optimal signal is then pieced together by joining these piece-wise optimal control solutions. This concept has been explored resulting in the well-established method of sequential quadratic programming (see [22, 23] and references therein). However, it must be reminded that there are catches inherent in this methodology when applied to engine transient control problems. Indeed, the original optimization problem applies to piecewise static processes while for engine transient control, it is exactly the time duration to be "optimized" and the philosophy of partition over time only leads to sub-optimal solutions other than global optimization.

To fully comprehend the above extremely important "non-optimality nature" of all the optimization-yet-partition-based methods (unfortunately this point is somehow under-estimated or even misunderstood in aeroengine control design practice). Consider an acceleration process from idle $n_i$ to TO $n_{\max}$, the acceleration time is:

$$t_{ac} = \left(\frac{\pi}{30}\right)^2 I \int_{n_i}^{n_{\max}} \frac{ndn}{W_a(\omega_T \eta_m - \omega_c)} \tag{5.1}$$

where: $\omega_T$ is the power produced by turbines in unit of kw/kg; $\omega_c$ is the power consumed by compression systems; $\eta_m$ is the mechanical efficiency; $I$ is shaft inertia in unit of kg.m$^2$. It is seen clearly that the optimization of $t_{ac}$ depends on rotational speed which leaves the integral indefinite while $\omega_T$ renders the corresponding optimization problem non-determinable as it is determined by turbine expansion ratio $\pi_T^*$ and turbine inlet temperature $T_4^*$ as:

$$\omega_T = c_p' T_4^* \left(1 - \frac{1}{\pi_T^{*\frac{k'-1}{k'}}}\right) \eta_T^* \tag{5.2}$$

In practice, the optimal control for acceleration is obtained by formulating the following optimization problem:

$$\min J = \int_{t_0}^{t'_f} (n_H - n_{H\max})^2 dt \tag{5.3}$$

where: $t'_f$ is a pre-specified value with $t'_f \geq$ tf, and $t_f$ is the acceleration time, which is itself a varying parameter determined by high-pressure shaft speed $n_H$ reaching to maximum. To make the above optimization problem amenable to algorithmic solvers, it is transformed into a discrete quadratic representation:

$$\min J = (1 - \frac{n_H(k)}{n_{H\max}})^2 \Delta t \tag{5.4}$$

where: $\Delta t$ is chosen typically to be the engine electronic controller sampling rate as 0.02 s. Such an optimization problem is solvable. However, it must be pointed out that the resulting solution is sub-optimal due to:

(1) The original transition time optimization is transformed into (discretized) shaft speed optimization, and the resulting near-optimal solution is only guaranteed through engine propulsion principle consideration;
(2) The partition of transition time into discrete constant $\Delta t$ is only an approximation;
(3) Optimization of (4) requires invoking of linearized engine model over each sampling period. Although the model is actually "extracted" from the nonlinear component-level model, linearization implies that the resulting model is essentially feasible around steady state conditions. While engine transient control refers to transitional and dynamic states, such an invoking lead naturally to a sub-optimal solution.

Thus, it is seen that the partition-based optimal transient control is inherently sub-optimal. The utilization and application to large envelope control design must be carefully exercised. The dissemination is delineated in the following sections. The structure of this chapter is organized as follows: Sect. 5.1 presents a sequential quadratic optimization-based transient control method; Sect. 5.2 introduces another optimization method of augmented supporting machine for aeroengine transient and limit protection control integration. Section 5.3 then provides a comparative study for the optimization-based designs before a summary is given in Sect. 5.4.

## 5.1 Sequential Quadratic Optimization-Based Transient and Limit Protection Control

### 5.1.1 Performance Index Selection

To begin with, the conventional performance index (3) is augmented to include turbine inlet temperature as:

$$\min J = \frac{1}{2}\int_{t_0}^{t'_f} (n_H - n_{H\max})^2 dt + \frac{1}{2}\int_{t_0}^{t'_f} (T_4 - T_{4\max})^2 dt \tag{5.5}$$

where: $t_0$ is the initial time while $t_f$ is the (unknown) ending time with $t'_f$ as predesignated computational time constant ($t'_f > t_f$). The augmentation is for limit protection by making "penalty" to $T_4$, which is different from previous research solely for consideration of acceleration in $n_H$. The corresponding control variables are chosen to be fuel flow $W_f$ and nozzle area $A_8$. Now it is ready to discretize the performance index as:

$$\min J = 0.5 \times \left(1 - \frac{n_H(k)}{n_{H\max}}\right)^2 \Delta t + 0.5 \times \left(1 - \frac{T_4(k)}{T_{4\max}}\right)^2 \Delta t \tag{5.6}$$

where: $k = 1, 2, \ldots N$, as positive natural numbers denoting the number of divisions over acceleration process, while $\Delta t$ is fixed to be 0.02 s.

From the propulsion principle, it is known that the more the fuel injects to the burner, the higher the turbine inlet temperature is; and consequently, the larger power the turbines can produce. The extra surplus power certainly provides good engine acceleration property. The performance index above is chosen for high-pressure rotational speed to reach its maximum while retaining the maximal turbine inlet temperature and this can fully take advantage of engine potential. Should this strategy be implemented over the control frame from $t_0$ to $t'_f$, it is expected to result in fast acceleration. Albeit a suboptimal one is obtained; it is still desirable so long as the performance can be improved over conventional designs.

### 5.1.2 Derivation of Optimal Operation

The following key aspects are usually involved in obtaining the solution:

(1) Engine Model: given flight conditions for height $H$ and Mach number $Ma$, engine states are determined by control variables. This is expressed as:

$$X = F(u) \tag{5.7}$$

where: $X = \{F, n_H, n_L, T_4, SM_c, SM_f, \ldots\}$ as engine states and performance variables; while $u = \{A_8, W_f\}$ is the control variables.

(2) Performance Index: choose the performance index such as (5.5) with discrete counterpart (5.6).

(3) Determination of Initial Conditions: the initial conditions of height and velocity as well as idle speed need to be specified, e.g. $n_{H,idle} = 0.86 n_{H,\max}$. Situations over flight envelope (varying height and velocity) are investigated in the next section.

(4) Determination of Control Variables: fuel flow $W_f$ and nozzle area $A_8$ are usually chosen as control variables, although other options exist with minor adjustment.

(5) Determination of Constraints: the resulting nonlinear optimization problem is constrained by engine protection requirements such as over-speed, over-temperature, surge and stall preventions etc.

Obviously, constraints have significant influence upon the resulting solutions. And for the case of acceleration, the most often encountered ones should include:

*Over-speed Protection for High-pressure Rotational Speed*:

$$\max(n_H) \le n_{H\max} \tag{5.8.1}$$

*Over-speed Protection for Low-pressure Rotational Speed*:

$$\max(n_L) \le n_{L\max} \tag{5.8.2}$$

*Turbine Inlet Over-temperature Protection*:

$$\max(T_4) \le T_{4\max} \tag{5.8.3}$$

*Burner Over-fuel Protection*:

$$\max(far) \le far_{\max} \tag{5.8.4}$$

*High-pressure Compressor Surge Protection*:

$$\min(SM_c) \ge SM_{c\min} \tag{5.8.5}$$

*Low-pressure Compressor Surge Protection*:

$$\min(SM_f) \ge SM_{f\min} \tag{5.8.6}$$

where in (5.8.4), far is the burner fuel to air ratio defined as: $far = W_f/m_a$, and $m_a$ is the air flow through the burner. Meanwhile, besides the above protection requirements, the fuel injection devices and nozzle control devices are also subject to physical constraints, and these should also be taken into account.

*Range of Burner Fuel Injection*:

$$W_{f,idle} \le Wf \le W_{f\max} \tag{5.8.7}$$

*Range of Nozzle Area Actuation*:

$$A_{8\min} \le A_8 \le A_{8\max} \tag{5.8.8}$$

*Range of Burner Fuel Injection Ratio*:

$$\max(\Delta W_f) \le \Delta W_{f\max} \tag{5.8.9}$$

*Range of Nozzle Area Actuation Ratio*:

$$\max(\Delta A_8) \le \Delta A_{8\max} \tag{5.8.10}$$

The above constraints are considered in the optimization process with unity unification to avoid large parameter value derivation as follows:

$$\mathrm{c1}=\frac{\mathrm{n}_H(\mathrm{k})}{n_{H\max}} - 1 \tag{5.9.1}$$

$$\mathrm{c2} = \frac{\mathrm{n}_L(k)}{n_{L\max}} - 1 \tag{5.9.2}$$

$$\mathrm{c3} = \frac{\mathrm{far}(k)}{\mathrm{far}_{\max}} - 1 \tag{5.9.3}$$

$$\mathrm{c4} = \frac{T_4(k)}{T_{4\max}} - 1 \tag{5.9.4}$$

$$\mathrm{c5} = 1 - \frac{SM_c(k)}{SM_{c\min}} \tag{5.9.5}$$

$$\mathrm{c6} = 1 - \frac{SM_f(k)}{SM_{f\min}} \tag{5.9.6}$$

$$\mathrm{c7}=\frac{|\Delta W_f|}{\Delta W_{f\max}} - 1 \tag{5.9.7}$$

$$c8 = \frac{|\Delta A_8|}{\Delta A_{8\ \max}} - 1 \tag{5.9.8}$$

These aspects will be illustrated with numerical examples and solutions in the following sections.

### 5.1.3 Optimal Transient and Limit Protection Control: Numerical Study

#### 5.1.3.1 Optimal Acceleration Control

Now first consider the acceleration process, it is well-known that change of fuel injection and nozzle actuation can lead to fast change of performance variables to reach limits. Consequently, a variety of limits must be enforced to ensure safe operation over flight envelope. In present investigation, the following constraints are put forward:

*Range of Burner Fuel Injection*: 0.45 kg/s $\leq W_f \leq$ 1.144 kg/s;

*Range of Nozzle Area Actuation*: 0.2435 $m^2 \leq A_8 \leq$ 0.2935 $m^2$;

*Range of Burner Fuel Injection Ratio*: $\Delta W_{f\ \max} = (0.4\ \text{kg/s})/\text{s}$;

*Range of Nozzle Area Actuation Ratio*: $\Delta A_{8\ \max} \leq (0.06\ \text{m}^2)/\text{s}$;

*High-pressure Maximum Rotational Speed*: $n_{H\ \max} = 12794rpm$;

*Low-pressure Maximum Rotational Speed*: $n_{L\ \max} = 9927rpm$;

*Maximum Turbine Inlet Temperature:* $T_{4\ \max} = 1640$ K;

*Burner Over-fuel Protection Boundary*: $far_{\max} = 0.022$;

*Minimum High-pressure Compressor Surge Margin*: $SM_{c\ \min} = 0.115$;

*Minimum Low-pressure Compressor Surge Margin*: $SM_{f\ \min} = 0.32$;

**Remark 1** The boundaries are generated from component-level simulation reaching their corresponding limits as constraints. Thus, the bounding values are different from those actually implemented in FADEC. But such handling does not repudiate the feasibility of the methodology.

**Remark 2** It is worth pointing out that turbine inlet temperature is limited above by 1640 K, which is larger than the maximum value 1630 K. This is allowable since engine acceleration time is short allowing temporary over-limit. And this is a common strategy for performance seeking control.

Now the SQP algorithm is implemented in the environment of Matlab2014 software platform, where the step size is fixed to be 0.02 s, mimicking a typical FADEC EEC sampling rate. The results from idle to TO for $H = 0$ and $Ma = 0$ are shown in Fig. 5.2, also shown are the results for conventional schedule-based acceleration method.

From Fig. 5.2a, it is seen that the fuel flow curve consists of four segments with each segment leveling off to reach maximum; nozzle changes area from maximum to minimum with max-slope and then maintains at minimum area (Fig. 5.2b); meanwhile both high-pressure and low-pressure rotational speeds reach their max-values from idle speed (Fig. 5.2c, d). Figure 5.2e shows the turbine inlet temperature reaching its protection value 1640 K, maintaining there while decreasing gradually to maximum value 1630 K; a similar trend applies to fuel-to-air ratio in Fig. 5.2f. Compressor surge margin quickly decreases to minimum and gradually increases back to an optimal value, while at the same time, fan surge margin keeps decreasing to minimum maintaining constant till the end of acceleration.

It is thus seen clearly that during the first stage of acceleration, fuel injection to the burner keeps increasing while limited by maximum injection ratio, and this leads to fast increase in rotational speeds, while quick decrease in surge margins. When surge margin decreases to the minimum, fuel flow is actually limited by surge margins constraints with gradual increase during the second stage. In the third stage, turbine inlet temperature reaches protection value with continuous increase in fuel flow (but fuel flow rate is smaller than previous stage), and surge margins keep increasing to stabilizing values. At the final stage, fuel flow reaches its maximum and rotational speeds keep increasing to required values.

In summary, the whole acceleration process takes 3.3 s, and in comparison with conventional schedule-based acceleration method, the SQP approach results in fast acceleration and is thus advantageous.

#### 5.1.3.2 Optimal Deceleration Control

Now consider the deceleration process where the pilot usually manipulates the PLA from max-position to idle angle. This leads to quick decrease in fuel flow, which very likely causes burner flame-out due to the fact that engine still works at very high rotational speed. Protection must be made to prevent lean blow-out (this is especially true when combustion condition deteriorates over high attitude). Now a performance index can be defined as below:

$$\min J = \frac{1}{2}\int_{t_0}^{t'_f} (n_H - n_{H\min})^2 dt + \frac{1}{2}\int_{t_0}^{t'_f} (T_4 - T_{4\min})^2 dt \tag{5.10}$$

Discretize the index:

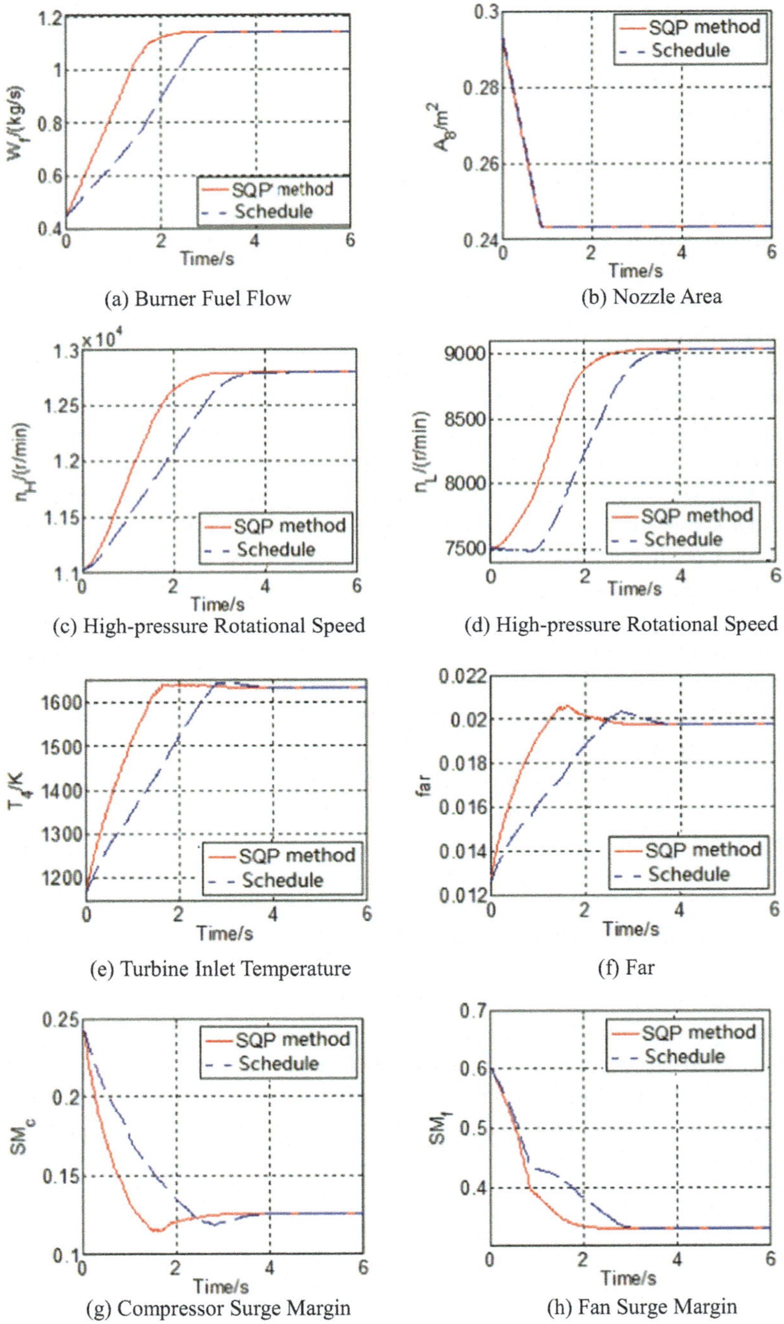


(a) Burner Fuel Flow

(b) Nozzle Area

(c) High-pressure Rotational Speed

(d) High-pressure Rotational Speed

(e) Turbine Inlet Temperature

(f) Far

(g) Compressor Surge Margin

(h) Fan Surge Margin

**Fig. 5.2** Acceleration control with SQP optimization and conventional schedule-based method

$$\min J = 0.5 \times \left(1 - \frac{n_H(k)}{n_{H\min}}\right)^2 \Delta t + 0.5 \times \left(1 - \frac{T_4(k)}{T_{4\min}}\right)^2 \Delta t \tag{5.11}$$

with the following constraints:

*Range of Burner Fuel Injection*: $0.45kg/s \le W_f \le 1.144$ kg/s;

*Range of Nozzle Area Actuation*: $0.2435\ \mathrm{m}^2 \le A_8 \le 0.2935\ \mathrm{m}^2$;

*Range of Burner Fuel Injection Ratio*: $\Delta W_{f\ \max} = (0.4\ \mathrm{kg/s})/\mathrm{s}$;

*Range of Nozzle Area Actuation Ratio*: $\Delta A_{8\ \max} \le (0.06\ \mathrm{m}^2)/\mathrm{s}$;

*High-pressure Idle Rotational Speed*: $n_H = 10480rpm$;

*Low-pressure Idle Rotational Speed*: $n_L = 6759rpm$;

*Turbine Inlet Temperature:* $T_4 = 1058$ K;

*Burner Lean Blow-out Boundary*: $far_{\min} = 0.0125$;

**Remark 3** Constraints for fan and compressor surge margins do not come into force in principle. However, when an engine has compressors with variable guide vanes or bleeding valves, these adjustable components must have a changing rate large enough to accommodate fast decrease in rotational speeds, otherwise the decrease rate for fuel flow must be limited.

With these preliminaries and the same environmental setup as acceleration design, the optimal results for deceleration can be obtained in Fig. 5.3:

It is seen from Fig. 5.3a that fuel flow consists of two stages: one from start of deceleration to 0.94 s where fuel flow quickly decreases; the other one from 0.94 to 3.85 s where the curves obviously level off. Also, during the first stage, the fuel flow rate keeps constant to be maximum, indicating that the constraint in maximum fuel flow rate is in force. During the second stage, the fuel-to-air ratio in Fig. 5.3f retains its constant yet minimum value, showing the fact that engine has been working at the boundary for lean blow-out. Hence the fuel flow levels off due to the influence from fuel-to-air ratio.

In summary, the deceleration process lasts for 3.85 s; and it starts with fuel flow remaining in maximum decreasing rate. Due to the inertia, the engine still works at high rotational state, and this causes quick increase of air flow entering into burner, leading to the fast reach to burner lean blow-out boundary, maintaining until idle state. This demonstrates the effectiveness of the method.

**Remark 4** However, it must be noted from Fig. 5.3c-e that there are steady-state errors in $n_H$ (4.7%), $T_4$ (9.168%) and $n_L$ (2.08%). This is due to the fact that fuel flow is essentially of open-loop nature. The corresponding steady-state errors can be reduced by augmenting the open-loop fuel flow injection with a closed-loop remedy.

**Remark 5** The optimal deceleration process is actually to decelerate the engine along the lean blow-out boundary while respecting to the related constraints. This "rule-of-thumb" is thus validated by the theoretical results above.

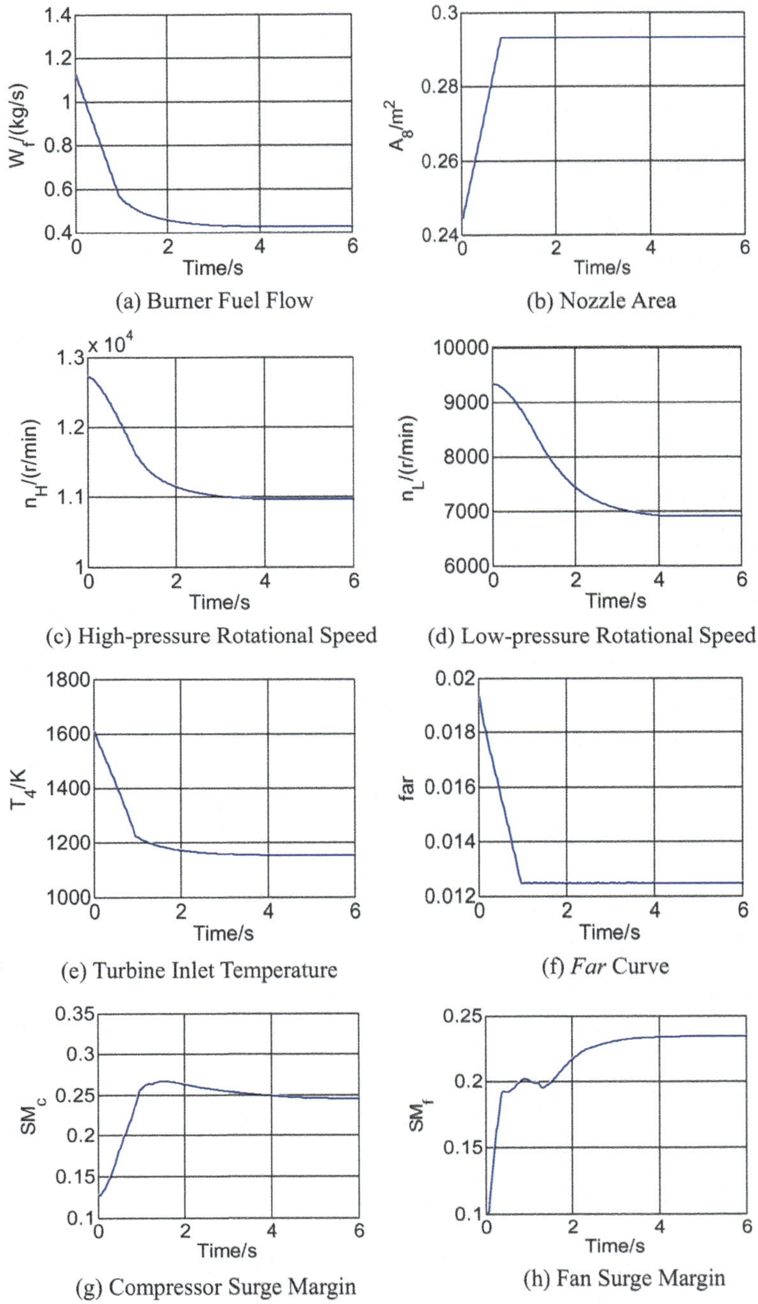


**Fig. 5.3** Deceleration control with SQP optimization

### 5.1.4 Control Integration Over Full Flight Envelope

Aeroengine transient and limit protection control is investigated in the above section for fixed flight conditions. It is thus important to consider the situation over full flight envelope. Usually, two methods can be utilized to obtain the control laws over flight envelope: similarity transformation method and the method of envelope division. Similarity transformation method utilizes the so-called "corrected parameters" to transform the transient control laws to the uniform flight condition of $H = 0$ and $Ma = 0$. For example, as engine acceleration is affected by flight and environmental conditions, engine structural and operational parameters, it is instructive to write Eq. (5.1) to be:

$$dt_{ac'cor} = \left(\frac{\pi}{30}\right)^2 I \frac{n_{cor} dn_{cor}}{N_{T,cor} - N_{c,cor})} \tag{5.12}$$

where: $n_{cor} \equiv \frac{n}{\sqrt{T_{t2}}}$, $N_{T,cor} \equiv \frac{N_T}{p_{t2}\sqrt{T_{t2}}}$, and $N_{C,cor} \equiv \frac{N_C}{p_{t2}\sqrt{T_{t2}}}$ are the corrected rotational speed, corrected turbine power, and corrected compressor power, respectively, with the corresponding corrected acceleration time $t_{ac,cor} \equiv t_{ac}\frac{p_{t2}}{\sqrt{T_{t2}}}$. Refer to (5.12), for steady-state operation, engine states will be similar with additional corrected fuel flow $\frac{W_f}{p_{t0}\sqrt{T_{t0}}}$ for transient operations. Therefore, control laws over flight envelope can be generated through those at $H = 0$ and $Ma = 0$ conditions.

However, one of the problems with this method is exactly due to failure of "similarity conditions". In this case, the envelope division method, namely, dividing the flight envelope by height and Mach numbers into a number of zones, transient control laws are then studied zone-by-zone before an integration strategy is applied. This method is of a trial-and-error nature, but is often taken as the practical approach to transient control design.

#### 5.1.4.1 Height Property

To proceed, it has been assumed that the engine under discussion has the envelope of $H = 0 \sim 4$ km and $Ma = 0 \sim 0.8$. Application of the optimization algorithms into conditions (1) $H = 0$, $Ma = 0$; (2) $H = 2000$, $Ma = 0$; and (3) $H = 4000$, $Ma = 0$. The following results for "height properties" are obtained as shown in Fig. 5.4:

From Fig. 5.4a, it is seen that acceleration is first constrained by maximum fuel flow rate and then by surge margins before turbine inlet temperature $T4$ enforces its limit. When height increases to $H = 2$ km (Fig. 5.4b), the trend retains similar but surge margins begin play their role ahead of time while pushing the enforcement of $T4$ much to higher rotational speeds. With height increasing to $H = 4$ km, surge margins even exercise their forces in such a way that $T4$ moves out of horizon. The crossing regions are drawn in Fig. 5.4d, indicating an acceleration line for different

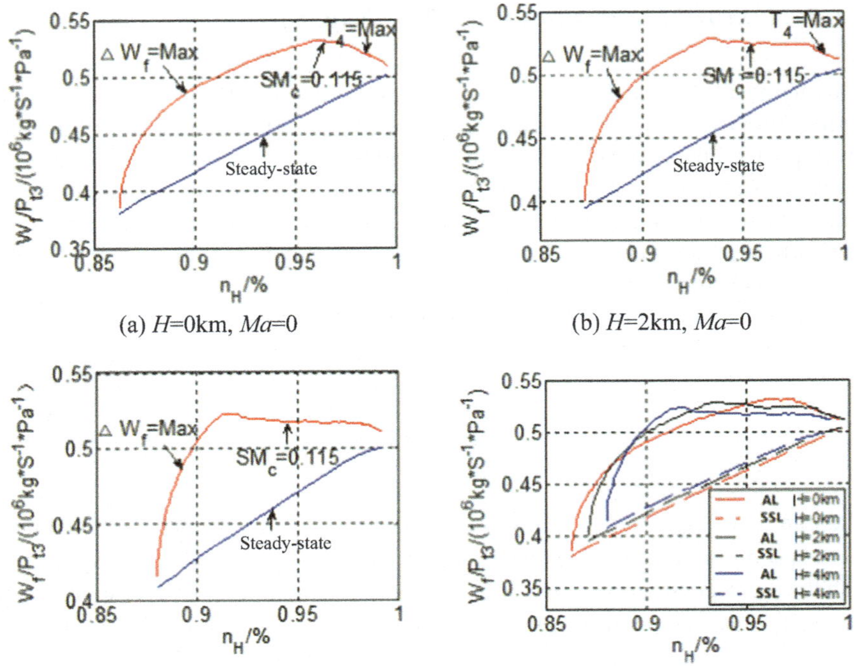


(a) $H$=0km, $Ma$=0

(b) $H$=2km, $Ma$=0

(c) $H$=4km, $Ma$=0

(d) Boundaries for Various Height

**Fig. 5.4** Height properties (AL: Acceleration Line; SSL: Steady-state Line)

velocity. The optimal acceleration line will be the uppermost curve within the crossing regions.

### 5.1.4.2 Velocity Property

Similarly, "velocity properties" can be obtained by considering the application of the optimization algorithms into following conditions (1) $H = 0$, $Ma = 0$; (2) $H = 0$, $Ma = 0.4$; and (3) $H = 0$, $Ma = 0.8$. This leads to the results in Fig. 5.5.

Figure 5.5b can be best read by comparing with Fig. 5.4b, where the roles of surge margins and turbine inlet temperature $T4$ are switched with increased $T4$ influence over increasing velocity. Finally, $T4$ becomes the dominating constraints in Fig. 5.5c. Again, the crossing region is also illustrated in (d) as the feasible acceleration line within the region.

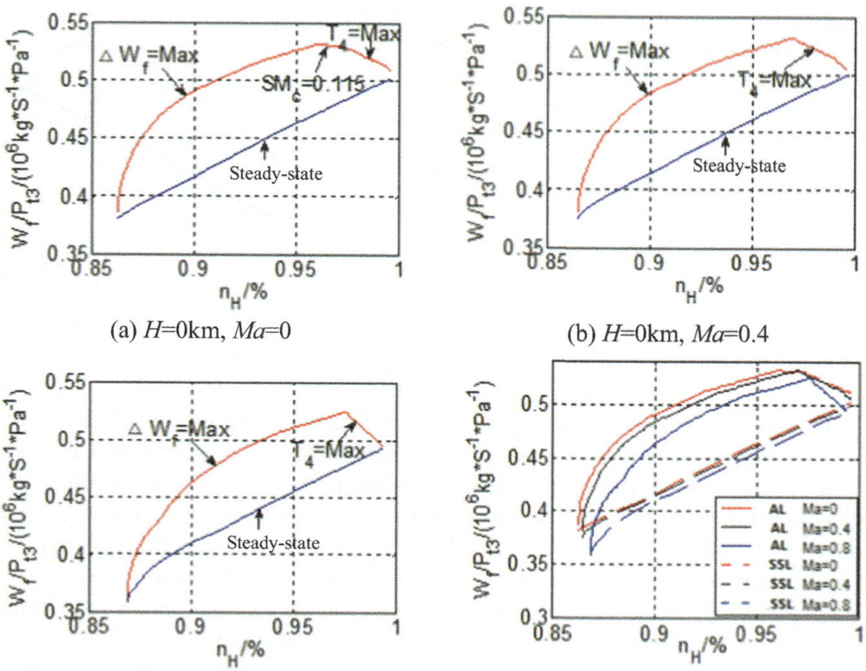


(a) $H$=0km, $Ma$=0

(b) $H$=0km, $Ma$=0.4

(a) $H$=0km, $Ma$=0.8

(d) Boundaries for Various Velocity

**Fig. 5.5** Velocity properties (AL: Acceleration Line; SSL: Steady-state Line)

#### 5.1.4.3 Optimal Operation

With both height and velocity properties, the boundaries can be plotted together as the control envelope. The crossing region is the feasible solutions for acceleration. Although any of the choice within the region can be taken as an optimal solution, it is those near the upper boundary that would achieve better acceleration property than other choices, see Fig. 5.6.

### *5.1.5 Conclusion & Discussions for SQP Method to Control Integration*

An SQP optimization-based method has been considered for transient control of turbofan engines with limit protection. Design of both acceleration and deceleration controls has also been given in details with specific constraints. While the method has been demonstrated to be effective for fixed flight conditions, full flight envelope design has also been considered.

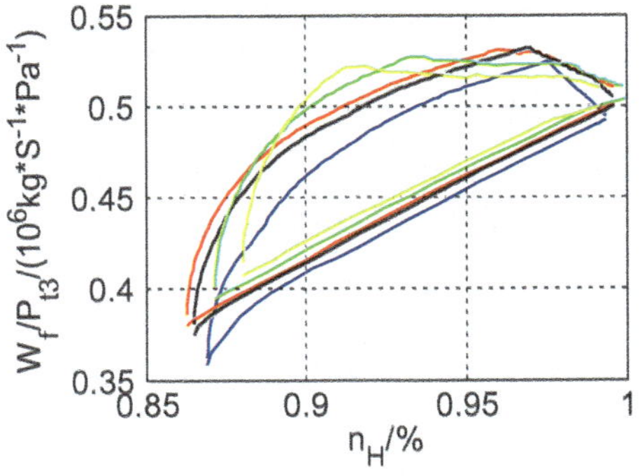


**Fig. 5.6** Optimal operation by integrating height and velocity properties

One of the drawbacks of the proposed methodology is the sub-optimality of the resulting solutions, due to discretization of performance index as well as utilization of linearized steady-state models during transition processes. However, it has also been shown that the proposed method results in better acceleration properties than conventional schedule-based methods. Indeed, due to environmental disturbances, model uncertainties, and component properties drifts over engine operations, it is very rare to be able to obtain accurate models as well as estimation of optimal performance boundaries. As a consequence, it will be very useful for a methodology to provide design guidance. The proposed method does result in a series of results suggestive to practical implementations.

## 5.2 Active Set Method for Transient and Limit Protection Control

To achieve limit protection control with large envelope operation, Sect. 5.1 has demonstrated the power of utilizing nonlinear programming techniques [24–26]. Theoretically dynamic programming, particularly quadratic programming problems can be solved by a variety of methods (see the review [27]). An interested one is presented here as active set method (ASM), which is also one of the interior-point algorithms [28]. Indeed, from an algorithmic perspective, should an appropriate performance index be defined, optimal solution set can be reached by all the optimization methods. The differences among them can be speed of convergence, occupation of computational resources etc. From limit protection control perspective, however, ASM has the property that its solution is always feasible during each iteration, while the search direction is defined based on current effective constraint set. This feature fits well for engine limit protection control where all the limits are required to be satisfied at all time. Thus, even any breakdown or overrun ending at intermediate solutions, the ASM algorithm can still prevent from exceeding limits. This is the reason that ASM is adopted in this paper, besides the fact that it has not been investigated for engine control applications. These issues are organized in the following sections: Sect. 5.2.1 provides a brief introduction to active set method, while analysing the problem of engine performance optimization with this method; Sect. 5.2.2 presents the optimization results with robustness analysis; to demonstrate the effectiveness of the proposed method, a comparison with a nonlinear minimum variance based method is carried out in Sect. 5.2.3; finally concluding remarks are presented in Sect. 5.2.4.

### *5.2.1 Nonlinear Control with Active Set Optimization*

Consider a generic constrained quadratic programming problem:

$$\begin{aligned} \min \quad & Q(x) = \frac{1}{2}x^T G x + g^T x \\ \text{s.t.} \quad & Ax \geq b \end{aligned} \tag{5.13}$$

where: $Q$ is a positive definite matrix of rank $n$; $g \in R^n$, $b \in R^m$, and $A \in R^{m\times n}$ with $rank\ A = m$. The key to ASM algorithm is to treat one of the inequality constraints as equality while ignoring other inactive constraints for performance index optimization; then a new feasible solution is obtained before the procedure is iterated. By this way, the problem is transformed into a quadratic convex programming with only equality constraints such as:

$$\begin{aligned} \min \quad & f(x) \\ \text{s.t.} \quad & a_i^T x = b_i, i \in I_k \end{aligned} \tag{5.14}$$

At $k^{th}$ iteration, move the origin to $x_k$ and define: $d_k = x - x_k$, the (2) transforms to:

$$\begin{aligned} f(x) &= \frac{1}{2}(d_k + d_k)^T Q(d_k + x_k) + g^T(d_k + x_k) \\ &= \frac{1}{2}d_k^T Q d_k + d_k^T Q d_k + \frac{1}{2}x_k^T Q x_k + g^T d^T + g^T x_k \\ &= \frac{1}{2}d_k^T Q d_k + \nabla f(x_k)^T d_k + f(x_k) \end{aligned} \tag{5.15}$$

and the quadratic programming problem changes to another one with respect to $d_k$:

$$\begin{aligned} \min \quad & \frac{1}{2}d_k^T Q d_k + \nabla f(x_k)^T d_k \\ \text{s.t.} \quad & a_i^T d_k = 0, i \in I_k \end{aligned} \tag{5.16}$$

Then the following cases can be considered:

(1) $x_k + d_k$ is feasible and $d_k \neq 0$, then for $k+1$ iteration, take $x_{k+1} = x_k + d_k$;
(2) $x_k + d_k$ is non-feasible, then define $x_{k+1} = x_k + \lambda_k d_k$ with $\lambda_k$ as step size. To fulfill feasibility requirement, $\lambda_k$ is such chosen as for any $i \notin I_k$:

$$a_i^T(x_k + \lambda_k d_k) \geq b_i \tag{5.17}$$

As $x_k$ is feasible, $a_i^T x_k \geq b_i$, hence when $a_i^T d_k \geq 0$, Eq. (5.17) always hold for any non-negative $\lambda_k$; if $a_i^T d_k < 0$, a choice of positive number with:

$$\lambda_k \geq \min\left\{\frac{b_i - a_i^T x_k}{a_i^T d_k} | i \notin I_k, a_i^T d_k < 0\right\} \tag{5.18}$$

Equation (5.17) holds for any $i \notin I_k$. A better choice can be made for $\lambda_k = \min\left\{1, \hat{\lambda}_k\right\}$, where $\hat{\lambda}_k$ is the right-hand side term of (5.18). Consequently, if there exists $p \notin I_k$ such that:

$$\lambda_k = \frac{b_p - a_p^T x_k}{a_p^T d_k} < 1 \tag{5.19}$$

then at $x_{k+1}$:

$$a_p^T x_{k+1} = a_p^T(x_k + \lambda_k d_k) = b_p \tag{5.20}$$

This indicates that $a_p^T d \geq b_p$ is a feasible constraint at $x_{k+1}$. Therefore, a new feasible set is augmented for $I_{k+1}$, or $I_{k+1} = I_k \cup \{p\}$.

(3) $d_k = 0$, then $x_k$ is the optimal solution. Write $v_i^{(k)}$ as the Lagrangian multiplier of active set constraints with $i \in I_k$. Thus if $v_i^{(k)} \geq 0$ and $i \in I_k$, $x_k$ is designated to be optimal solution; if $q \in I_k$ exists such that $v_q^{(k)} < 0$, $x_k$ is not optimal and a new feasible set is defined to be $I_{k+1} = I_k \backslash \{r\}$, where $v_r^{(k)} = \min\left\{v_i^{(k)} | i \in I_k\right\}$.

Thus, the ASM algorithm runs as:

*Step 1*: for initial value $x_0$, choose the feasible active constraint set $I_0$, and let $k = 0$.

*Step 2*: solve the following optimization problem with equality constraint to obtain $d_k$:

$$\begin{aligned} \min \quad & \frac{1}{2} d_k^T Q d_k + \nabla f(x_k)^T d_k \\ \text{s.t.} \quad & a_k^T d_k = 0, i \in I_k \end{aligned}$$

if $d_k$=0, turn to Step 4; otherwise let $\lambda_k = \min\left\{1, \hat{\lambda}_k\right\}$, $x_{k+1} = x_k + \lambda_k d_k$, and calculate $\nabla f(x_{k+1})$, then turn to Step 3.

*Step 3*: if $\lambda_k < 1$, then define $I_{k+1} = I_k \cup \{p\}$, $k = k + 1$, and return to Step 2; if $\lambda_k = 1$, then define $k = k + 1$, and go to Step 4.

*Step 4*: Calculate the Lagrangian multiplier $v_i^{(k)}$, $i \in I_k$, then write $v_r^{(k)} = \min\left\{v_i^{(k)} | i \in I_k\right\}$. If $v_r^{(k)} \geq 0$, then stop the iteration and $x_k$ is the optimal solution; otherwise delete $r$ from $I_k$ and return to Step 2.

This is the general procedure for ASM algorithm. However, application to engine performance optimization is not straightforward, due to the special dynamical properties and the way of "recalling" engine model for computation. Care must be taken of and this is turned to in the next section.

#### 5.2.1.1 Engine Performance Optimization with ASM

The basic structure of the ASM optimization can be illustrated in Fig. 5.7. As for typical engine control systems, control variables are defined to be fuel flow $W_f$ and nozzle area $A_8$ while controlled variables can be high-pressure (HP) spool speed $n_H$, low-pressure (LP) spool speed $n_L$, turbine inlet temperature $T_4$, fuel-to-air ratio $far$, as well as two surge margins (SM) for fan $SM_f$ and compressor $SM_c$. Mathematically, an engine control system can be represented as:

$$X = F(U) \tag{5.21}$$

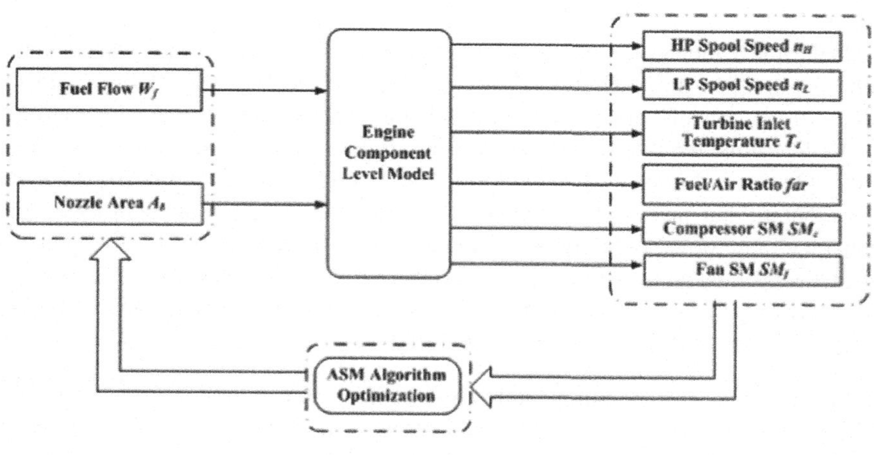


**Fig. 5.7** Engine Performance Optimization with ASM: Control Variables and Controlled Variables

where: $X$ denotes the set of state variables as $X = [n_H, n_L, T_4, SM_c, SM_f, far, sfc]$, and these are important parameters to characterize engine operations; $U$ is the set for control variables as $U = [W_f, A_8]$, and they are the control inputs to change the state of operation; $F$ represents a nonlinear function characterizing engine dynamics, which is essentially a component-level model. To apply the ASM algorithm, performance index and constraints must be designed.

#### 5.2.1.2 Performance Index Design

Selection of performance index has significant impact on optimization result, and its design must take engine dynamical properties into account. For engine acceleration and deceleration, it is the transition time (hence speed of regulation) that should be considered. As a consequence, the following performance indices can be potential candidates:

$$A.f = t_f - t_0 \tag{5.22}$$

$$B.f = \int_{t_0}^{t_f} (n_H - n_{H.aim})^2 dt \tag{5.23}$$

$$C.f \int_{t_0}^{t_f} (n_H - n_{H.aim})^2 dt \int_{t_0}^{t_f} (T_4 - T_{4.aim})^2 dt \tag{5.24}$$

In the above indices, $t_f$ is the acceleration/deceleration ending time, $t_0$ is the initial time, $n_{H.aim}$ is the desired high-pressure spool speed (e.g. TO speed, cruise speed, or idle speed etc.) while $T_{4.aim}$ is the corresponding desired turbine inlet temperature.

To determine the index, optimization algorithms must be brought into consideration. Quadratic forms are thus preferred due to convexity of the underlying optimization problem and implementation of corresponding computing algorithms. For both quadratic forms (5.22) and (5.23), index (5.24) further makes penalty upon $T_4$. From the propulsion principle, it is known that turbine inlet temperature $T_4$ directly influence the engine transition time, and thus is an important control variable for $t_f$. Henceforth, index (5.24) is chosen for the following investigations.

**Remark 1** It is worth pointing out that turbine inlet temperature $T_4$ has become one of the most important parameters to evaluate engine performance. In a certain sense, it becomes a metric for generation classification of modern turbofan engines.

However, the above performance indices are continuous functions and must be discretized:

$$\min \quad f(x) = \sum_{k=1}^{n} \left(1 - \frac{n_H(k)}{n_{H.aim}}\right)^2 \Delta t + \sum_{k=1}^{n} \left(1 - \frac{T_4(k)}{T_{4.aim}}\right)^2 \Delta t \tag{5.25}$$

A careful inspection of (5.25) reveals that step time $\Delta t$ should be provided, although its specification does not affect the implementation of programming algorithms. It is generally understood that $\Delta t$ is typically 20 ms for FADEC, and the optimization problem must be solved within this time frame. With these explanations, now (5.25) can be further simplified as:

$$\min \quad f(x) = \left[1 - \frac{n_H(k)}{n_{H.aim}}\right]^2 + \left[1 - \frac{T_4(k)}{T_{4.aim}}\right]^2 \tag{5.26}$$

where: $n_H(k)$ and $T_4(k)$ are the HP spool speed variable and turbine inlet temperature variable at $k^{th}$ step.

#### 5.2.1.3 Engine Constraints Design

Once the performance index is determined, constrains should be provided before accepting the programming algorithms. Usually the engine should be protected from exceeding the maximum HP spool speed, maximum LP spool speed, $T_4$ temperature limit, either fuel-rich blowout during acceleration or lean blowout for deceleration (usually defined as a safe area for fuel-to-air ratio $far$), as well as preventing from fan and compressor surges (defined as minimum surge margins $\Delta SM_f$ and $\Delta SM_c$). Besides these, some physical constraints should be considered such as fuel flow $W_f$ and nozzle area $A_8$ should only be varied within a pre-specified range.

In summary, the optimization problem is defined as follows:

$$\begin{aligned} \min \quad & f(x) = \left[1 - \frac{n_H(k)}{n_{H.aim}}\right]^2 + \left[1 - \frac{T_4(k)}{T_{4.aim}}\right]^2 \\ \text{s.t.} \quad & C(X) \in \varepsilon \end{aligned} \tag{5.27}$$

with: engine model $X = F(U)$ and $X = \left[n_H, n_L, T_4, \Delta SM_c, \Delta SM_f, far, sfc\right]$, $\varepsilon$ is the feasible set for constraints functions. A flow chart can be given to illustrate the optimization process in Fig. 5.8: the procedure starts with initialization for engine model to operate at desired state, which can be idle or any intermediate state above idle. For example, if optimal transition from idle to TO is required, the engine model runs to idle; that is, a steady-state control input $U = [W_f, A_8]$ is provided before the optimization procedure continues. At this state, the feasible set $I_0$ is defined and the algorithm then starts with an iterative solution to the corresponding optimization problems till a solution is obtained.

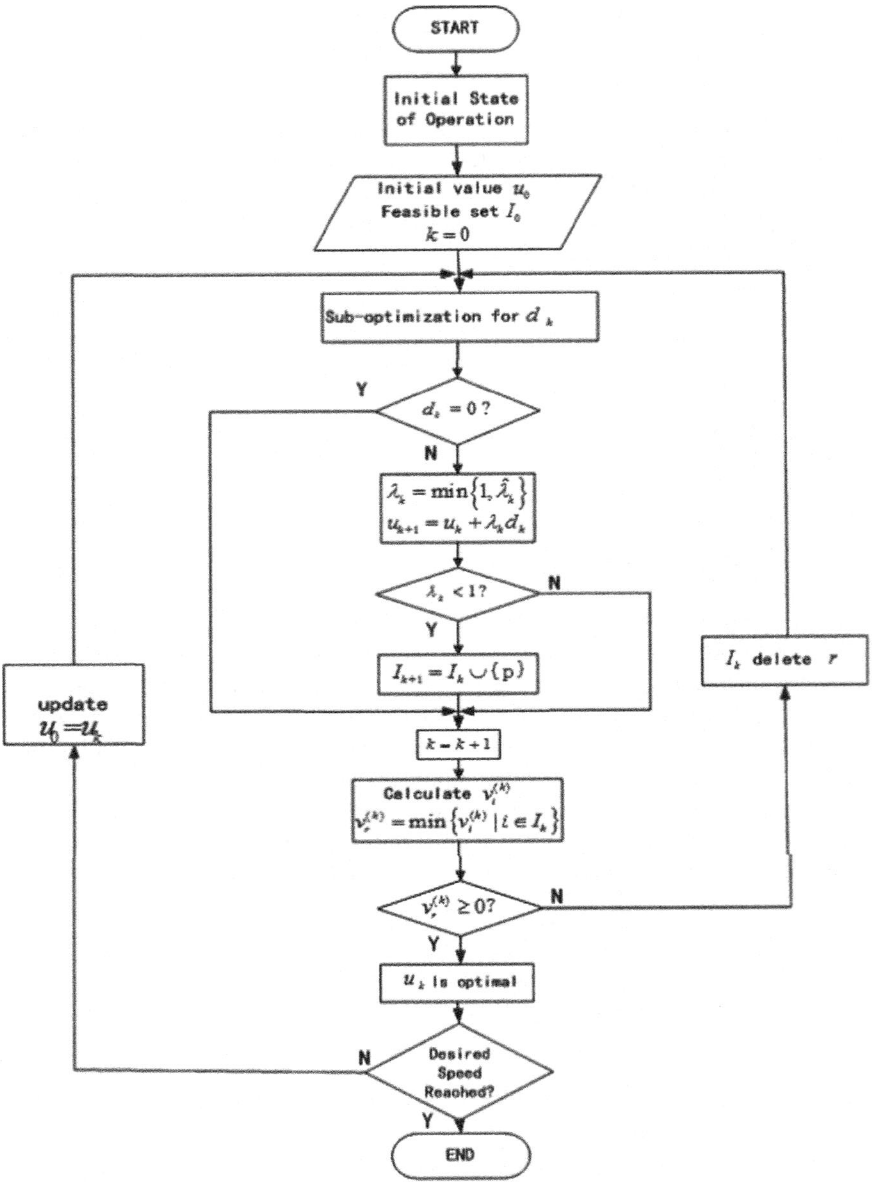


**Fig. 5.8** Flow chart for ASM-based engine optimization procedure

**Remark 2** The above development clearly reveals that there are many "catches" that should be taken care of when applying an established programming algorithm to engine control problems.

**Remark 3** One important point to be noted is that the resulting optimal solutions after the above "manipulations" are only sub-optimal to the original performance index (5.24). This is a generic issue associated with any optimization procedure due to the inherent property of engine transient operation. But from a practical design perspective, this would not be a concern should the resulting performance be improved, e.g. over conventional designs.

## 5.2.2 *Numerical Investigation: Analysis & Discussion*

### 5.2.2.1 Performance Optimization Results

To illustrate the above design procedures, first consider a typical acceleration process from engine idle to TO. For comparative reasons, two sets of conventional schedule-based acceleration and deceleration controls are also provided (Named as PLAN I and PLAN II). The following model parameters are provided based on a commercial, large by-pass turbofan engine component-level model, which has been developed for replication and validation of advanced control methodologies:

$$\min \quad f(x) = \left[1 - \frac{n_H(k)}{12794}\right]^2 + \left[1 - \frac{T_4(k)}{1640}\right]^2$$

$$\text{s.t.} \quad \begin{cases} 0.45\ \text{kg/s} \le W_f \le 1.144\ \text{kg/s}, & \Delta W_f \le (0.6\ \text{kg/s})/\text{s} \\ 0.2435\ \text{m}^2 \le A_8 \le 0.2935\ \text{m}^2, & \Delta A_8 \le \left(0.06\ \text{m}^2\right)/\text{s} \\ n_H \le 12794 rpm, & n_L \le 9027 rpm \\ T_4 \le 1640 K, & far \le 0.022 \\ \Delta SM_c \ge 30\%, & \Delta SM_f \ge 31\% \end{cases} \tag{5.28}$$

An application of ASM optimization leads to the following acceleration results in comparison with scheduled-based PLAN I and PLAN II as shown in Fig. 5.9.

It is seen from (a) that fuel flow increases very fast and then levels off due to the constraints from compressor surge margin and turbine inlet temperature; it gradually stabilizes reaching to its maximum allowable value. Nozzle area change (b) also complies with the practical rule during acceleration process, where nozzle area is set to be maximum for idle to reduce the turbine inlet temperature; and it changes to its minimum and remain there during engine acceleration. (c) and (d) are the spool speeds and it is seen that it takes 3 s for engine accelerating from idle to TO. Turbine inlet temperature $T_4$ in (e) quickly increases and gradually stabilizes at its maximum constrained value of 1640 K. (f) shows the thrust curve which is similar with spool speeds change, validating the fact that thrust is directly co-related

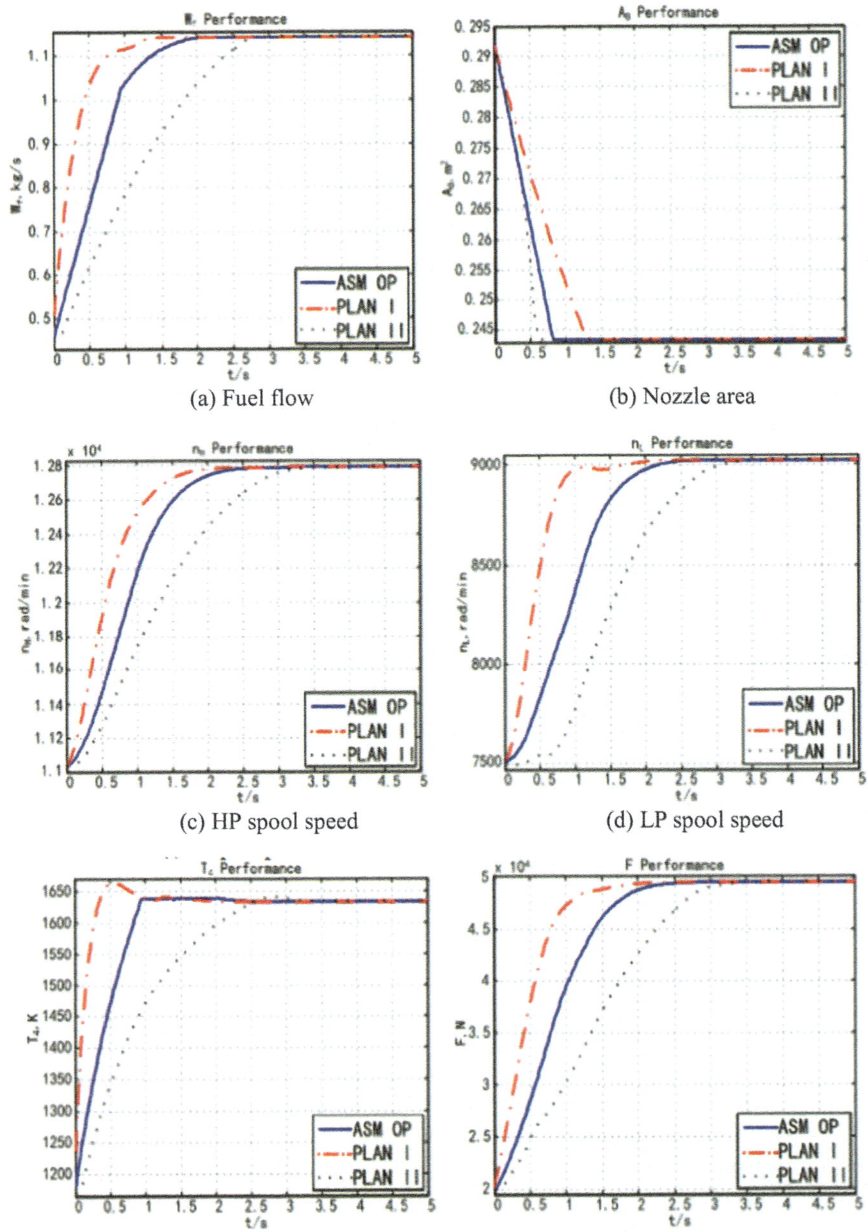


(a) Fuel flow (b) Nozzle area

(c) HP spool speed (d) LP spool speed

(e) Turbine inlet temperature (f) Thrust

**Fig. 5.9** Optimization results from idle to TO

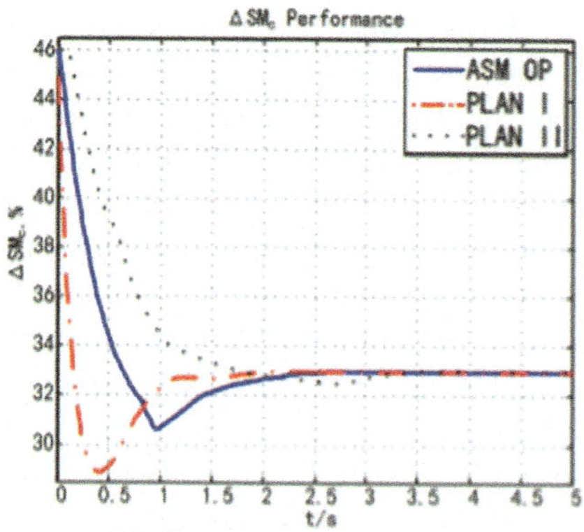


(g) Compressor surge margin

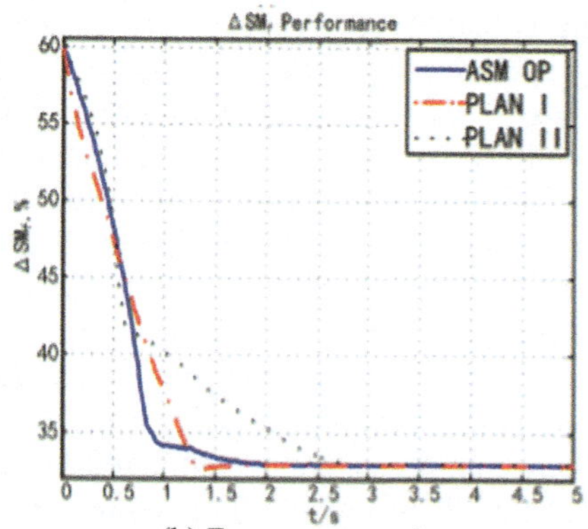


(h) Fan surge margin

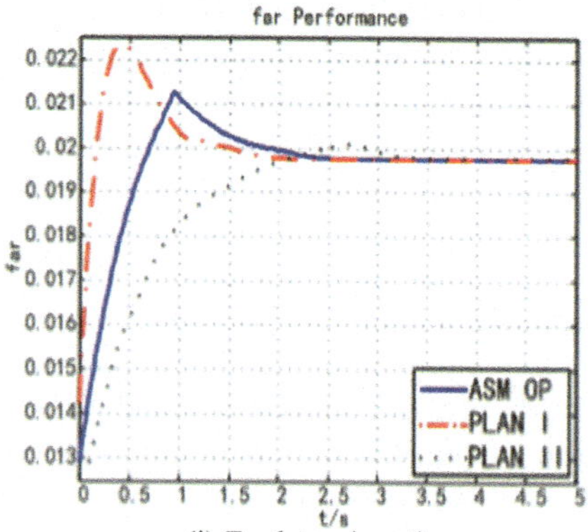


(i) Fuel-to-air ratio

**Fig. 5.9** (continued)

with engine rotational speeds. Compressor and fan surge margins in (g) and (h) quickly decrease and then obtain the authority to stabilize at the minimum values. Finally (i) shows the fuel-to-air ratio curve where it quickly increases to speed up the engine transition process; and then it gradually decreases from 1 to 2 s before final stabilization.

Thus the results show clearly that: fuel flow is first constrained by its (physically) maximum fuel injection ratio constraint during initial stage of acceleration; with spool speed increasing, surge margins decrease; when surge margins reach to their minimum allowable values, fuel flow is constrained by surge margins restriction; this then reduces the speed of acceleration and when speed increases further to approximate the boundary of maximum turbine inlet temperature, fuel flow gradually stabilizes to finally complete the acceleration process. This complies with the general physical understanding of the engine acceleration transitions.

In comparison with the optimization results with PLAN I and PLAN II, it is seen that ASM algorithm performs better than PLAN I and PLAN II in that PLAN I injects fuel faster while changing nozzle area slower than ASM; the trend reverses when comparing ASM with PLAN II. This makes PLAN I a bit aggressive while PLAN II over conserved. This is reflected from spool speeds (c) and (d), and further confirmed by (e). Indeed, the maximum turbine inlet temperature $T_4$ in PLAN I is actually higher than that in ASM, while slightly smaller in PLAN II. Due to the stringent requirement on turbine blades, temperature has great impact on blades life and even engine safety. Consequently, it could be a concern for the peak value in $T_4$ during 0–1 s in PLAN I. Meanwhile, PLAN II is too conservative from (f) where the thrust does not reach its maximum quickly. The aggressiveness of PLAN I and conservativeness of PLAN II are further confirmed by surge margins in (g) and (h) where compressor surge margin reaches below 30% in PLAN I while above 32% in PLAN II. This provides further "margin" for performance improvement.

Finally, with similar performance index and appropriate constraints defined the above optimization procedures can be carried out for deceleration process. Analogous analysis can be conducted that would provide useful design guidelines. But this will not be pursued here and an deceleration control is designed with consideration of another important issue in the next subsection—robustness of the optimization results.

#### 5.2.2.2 Robustness Analysis

Indeed, noises and disturbances are unavoidable and it is imperative to validate the robustness of the resulting control design. Now consider that $W_f$ and $A_8$ are varied over $0.45kg/s \leq W_f \leq 1.144$ kg/s and 0.2435 m$^2 \leq A_8 \leq 0.2935$ m$^2$, that is, the scale in fuel flow and nozzle area is 0.694 kg/s and 0.05 m$^2$. Now inject 5% and 10% noise in both fuel flow and nozzle area in a joint transient process first from idle to TO (acceleration) and then from TO to idle (deceleration). For ease of demonstration, only fuel flow and nozzle area as inputs, HP and LP spool speeds as outputs are presented as shown in Fig. 5.10.

It is seen that even for 10% noise level, both HP and LP spool speeds can maintain a tolerance of 0.3% at maximum values while there is an overshoot of around 1% for deceleration to idle speed for LP spool speed. Even though there exist some oscillations, the performance is acceptable for design specifications.

**Remark 4** To further evaluate the robustness of the algorithm, both 15% and 20% noise level have also been injected, and the results shows that HP spool speed at maximum has an overshoot of around 0.5% while an 1.5% variation is observed for LP spool speed. When the noise level reaches to 20%, the oscillations for HP and LP spool speeds are 0.69% and 1.9%, respectively. A 20% noise level obviously repudiates control system requirement.

**Remark 5** As a consequence, the ASM optimization possesses certain robustness and a tolerance of 15% noise level is achieved for proposed design. Of course, such an investigation is questionable due to the fact that noise and disturbance can emerge

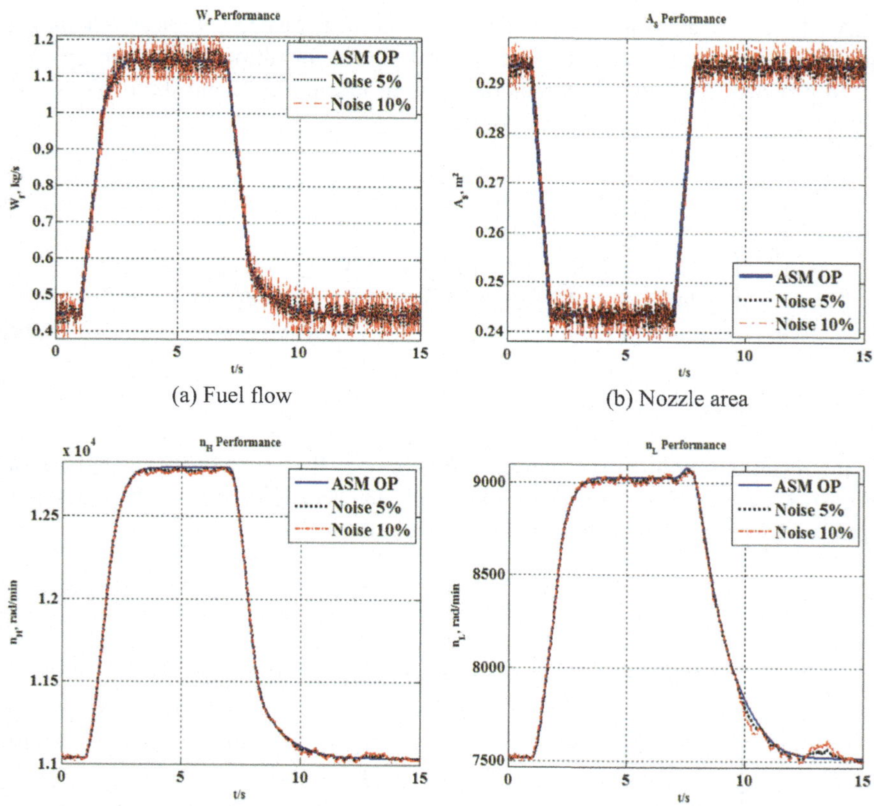


(a) Fuel flow

(b) Nozzle area

(c) HP spool speed

(d) LP spool speed

**Fig. 5.10** ASM optimization results with 5 and 10% noise injection

from many sources around control loops; their frequency and magnitude also vary significantly with engine operation conditions. But the simulation results do reveal the potential of ASM to control performance improvement.

### 5.2.3 *Active Set Method for Large Envelope Control: A Comparison*

To claim the effectiveness of above proposed method, it is imperative to make further evaluations. Specifically, two aspects should be considered for this purpose: (1) application to more elaborated scenarios over the envelope; and (2) comparison with other control design method. For (1), a new envelope is assumed to be acceleration from idle to 95% in HP spool speed; and then continue acceleration to 100%; this follows with a deceleration to 90%. For (2), conventional schedule-based design (PLAN I &

PLAN II) have been compared but in this section, an advanced nonlinear generalized minimum variance-based design method is designed for further evaluation.

To proceed, first consider acceleration from idle to 95%, and the performance index and constraints are designed to be:

$$\min \quad f(x)=\left[1-\frac{n_H(k)}{12154}\right]^2+\left[1-\frac{T_4(k)}{1460}\right]^2$$
$$\text{s.t.} \begin{cases} 0.45kg/s \le W_f \le 1.144 \text{ kg/s}, & \Delta W_f \le (0.6 \text{ kg/s})/\text{s} \\ 0.2435m^2 \le A_8 \le 0.2935 \text{ m}^2, & \Delta A_8 \le \left(0.06 \text{ m}^2\right)/\text{s} \\ n_H \le 12514rpm, & n_L \le 8711rpm \\ T_4 \le 1460 \text{ K}, & far \le 0.022 \\ \Delta SM_c \ge 30\%, & \Delta SM_f \ge 31\% \end{cases} \tag{5.29}$$

the resulting optimization stabilizes fuel flow at $W_f$=0.84789 kg/s and nozzle area at $A_8 = 0.2617$ m$^2$. Based on this result, the acceleration from 95 to 100% is optimized with performance index and constraints as:

$$\min \quad f(x)=\left[1-\frac{n_H(k)}{12794}\right]^2+\left[1-\frac{T_4(k)}{1640}\right]^2$$
$$\text{s.t.} \begin{cases} 0.45 \text{ kg/s} \le W_f \le 1.144 \text{ kg/s}, & \Delta W_f \le (0.6 \text{ kg/s})/\text{s} \\ 0.2435 \text{ m}^2 \le A_8 \le 0.2935 \text{ m}^2, & \Delta A_8 \le \left(0.06 \text{ m}^2\right)/\text{s} \\ n_H \le 12794rpm, & n_L \le 9027rpm \\ T_4 \le 1640 \text{ K}, & far \le 0.022 \\ \Delta SM_c \ge 30\%, & \Delta SM_f \ge 31\% \end{cases} \tag{5.30}$$

This provides the optimized results for fuel flow stabilized at $W_f$=1.144 kg/s and nozzle area at $A_8 = 0.2435$ m$^2$. Finally, deceleration from 100 to 90% can be designed with the corresponding performance index and constraints as:

$$\min \quad f(x)=\left[1-\frac{n_H(k)}{11514.6}\right]^2+\left[1-\frac{T_4(k)}{1284}\right]^2$$
$$\text{s.t.} \begin{cases} 0.45 \text{ kg/s} \le W_f \le 1.144 \text{ kg/s}, & \Delta W_f \le (0.6 \text{ kg/s})/\text{s} \\ 0.2435 \text{ m}^2 \le A_8 \le 0.2935 \text{ m}^2, & \Delta A_8 \le \left(0.06 \text{ m}^2\right)/\text{s} \\ n_H \ge 11514.6rpm, & n_L \ge 7728.69rpm \\ n_H \le 12794rpm, & n_L \le 9100rpm \\ T_4 \ge 1284 \text{ K}, & far \ge 0.0125 \end{cases} \tag{5.31}$$

This gives final results for fuel flow stabilized at $W_f$= 0.5869 kg/s and nozzle area at $A_8 = 0.2571$ m$^2$. Although not shown here, the results demonstrate that engine starts from idle at 4 s, and stabilizes at 9% at 8 s; then at 10 s it continues acceleration to 100%. The deceleration process then starts at 20 s, and stabilizes to

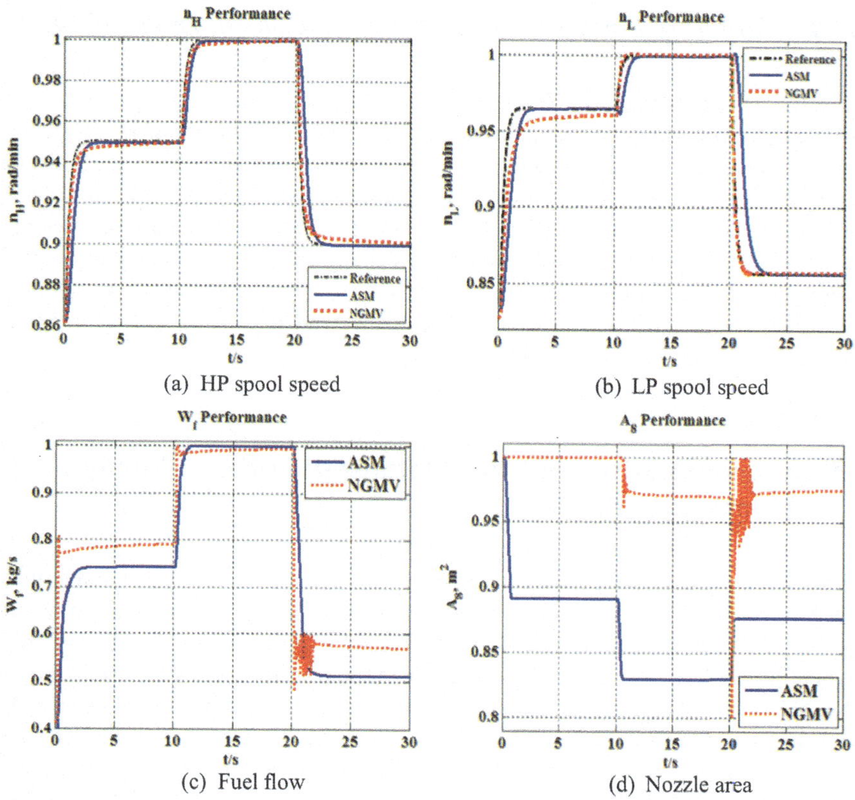


(a) HP spool speed (b) LP spool speed

(c) Fuel flow (d) Nozzle area

**Fig. 5.11** A Comparison of Performance between ASM and NGMV Control

90% at around 23 s. This large envelope operation is to be compared with a nonlinear generalized minimum variance-based (NGMV) method in Fig. 5.11.

From Fig. 5.11, it is seen that NGMV control has a longer regulation time than ASM and exhibits a 0.2% steady-state error for HP spool speed. Fuel consumption for NGMV is obviously higher than that of ASM, with oscillations present over deceleration. Nozzle area curve appears to have a similar trend with that of fuel consumption curve.

However, conclusive claims for NGMV control and ASM can not be made recklessly—the deviation for control inputs (fuel flow and nozzle area) is essentially due to the modeling errors where an identification model is utilized for design of NGMV control. A careful evaluation is to reach a conclusion that NGMV controller is easy to design and occupies less computational resources, while ASM possesses good control accuracy with small steady-state error. Besides, ASM solution always provides limit protection guarantee and NGMV control can only achieve the protection on an extensive trial-and-error basis.

### *5.2.4 Summary for Active Set Method for Control Integration*

An active set-based method has been introduced for performance optimization of turbofan engines. Detailed design procedures have been provided and the associated issues have been discussed to facilitate practical implementation. Numerical investigations have been presented to illustrate the design procedures as well as validating the effectiveness of the proposed method over conventional schedule-based designs. The superiority of the proposal has been further confirmed by a robustness analysis with injection of a variety of noise levels. Finally, a large envelope control scenario has been conducted with a comparative study with an advanced nonlinear generalized minimum variance-based control.

## 5.3 A Comparative Study for Optimization-Based Engine Transient Control

It is seen that optimization-based control is a powerful method to control integration. In recent years, a lot of research work on the engine of transition and limit protection control has been done [29–33]. Numerous studies are based on the algorithm of sequential quadratic programming to optimize the transition control variables as in Sect. 5.1. The research on the algorithm of active set method presented in Sect. 5.2 is rarely seen to optimize the engine control. Thus, this section aims to provide some experience on engine optimization design, namely comparing the two methods—active set method (ASM) [34] and sequential quadratic programming (SQP) [35] for optimization of acceleration and limit protection control. The investigation will be based on a two-spool mixed exhaust turbofan engine to be introduced below.

### *5.3.1 Aeroengine Model and Optimization Algorithms*

#### 5.3.1.1 Aeroengine Model

This paper uses some Component-level model for a Turbofan Engine [36] which base on the platform of Simulink to compare the two algorithms. Volume dynamic effect has been introduced to realize the dynamic simulation of engine. Therefore, the study of the transient and the set point control can be conducted. The model uses fuel flow and exhaust nozzle area as control inputs.

#### 5.3.1.2 Nonlinear Optimization Algorithms

The existence of the optimal solution can be verified by the limited amount of computation steps, it can also be obtained by means of numerical methods in a finite number of steps. Thus, to get the better method for SQP and ASM (both of them are quadratic programming algorithms) are compared. Abundant of research on the Nonlinear Optimization Algorithm has been taken. So, calculation steps are shown but not the proof. The problem can be abstracted as the following functions:

$$\min Q(x) = \frac{1}{2}x^T Gx + g^T x$$
$$s.t.\begin{cases} a_i^T x = b_i, i \in \varepsilon, \\ a_i^T x \geq b_i, i \in I. \end{cases} \tag{5.32}$$

where $G \in R^{n \times n}$ is symmetric positive definite, and its feasible region is $\Omega$.

***Framework of SQP is shown as the following***:

**STEP 1**: Set starting point $\mathrm{x}_0 \in R^n$. Set positive definite matrix $B_0$, and let $k = 0$.

**STEP 2**: Solve the following sub-problems to get $d_k$

$$\min f_k + \nabla f(x_k)^T d + \frac{1}{2}d^T B_K d$$
$$s.t.\begin{cases} c_i(x_k) + \nabla c_i(x_k)^T d = 0, i \in \varepsilon \\ c_i(x_k) + \nabla c_i(x_k)^T d \geq 0, i \in \mathrm{I}. \end{cases} \tag{5.33}$$

**STEP 3**: If $d_k = 0$, end.$\mathrm{x}_k$ is the K-T point of the original planning problem;

Else set $x_{k+1} = x_k + \alpha_k d_k$ where step length $\alpha_k \geq 0$ is based on some linear search.

**STEP 4**: Evaluate $B_k$ to get $B_{k+1}$ positive definite, set $k = k + 1$, go to STEP 2.

***Framework of ASM is shown as following:***

**STEP 1**: Set starting point $\mathrm{x}_0 \in R^n$, let $\mathrm{S}_0 = A(x_0)$,$k = 0$.

**STEP 2**: Calculate the following questions:

$$\min Q(x) = \frac{1}{2}(x_k + d)^T G(x_k + d) + g^T(x_k + d)$$
$$s.t.: a_i^T d = 0, i \in S_k \tag{5.34}$$

where $S_k \subset A(x_k)$ and $S_k$ is an estimate of $x_k$ for effective constraint index set $A(x_k)$.

Get the K-T point $(d_k, \lambda^k)$. If $d_k \neq 0$, go to STEP 3; Else, verify if $x_k$ is the K-T point of the following question:

$$\min Q(x) = \frac{1}{2}x^T Gx + g^T x$$
$$s.t. : a_i^T d = b_i, i \in S_k \tag{5.35}$$

If any $i \in S_k \cap I(x_k)$ gets $\lambda_i^k \geq 0$, then end the steps;

Else set $i_k = \arg \min\limits_{i \in S_k \cap I(x_k)} \{\lambda_i^k | \lambda_i^k < 0\}, S_{k+1} = S_k / \{i_k\}$ and let $x_{k+1} = x_k, k = k+1$ then go to STEP 3.

**STEP 3**: Set $x_{k+1} = x_k + \alpha_k d_k$.

**STEP 4**: If $\alpha_k \neq \hat{\alpha}_k$ let $S_{k+1} = S_k k = k + 1$ and go to STEP 2;

Else, if $i_k \notin S_k$ meet $a_{ik}^T(x_k + \alpha_k d_k) = b_{ik}$ let $S_{k+1} = S_k \cup \{i_k\} k = k + 1$ and go to STEP 2, where $\hat{\alpha}_k$ is the special estimates of $\alpha_k$.

## 5.3.2 Objective Function and Constraint Function

**Nomenclature**

| | |
|---|---|
| $N_h$ | Rotational speed of high-pressure spool |
| $N_l$ | Rotational speed of low-pressure spool |
| $N_{h_{\max}}$ | Max speed of $N_h$ |
| $N_{l_{\max}}$ | Max speed of $N_l$ |
| $T_4$ | Turbine forward temperature |
| $W_f$ | Fuel flow |
| $A_8$ | Nozzle area |
| $far$ | Fuel to air mass flow ratio |
| $smf$ | Surge margin of fan |
| $smc$ | Surge margin of compressor |

**Objective Function**

During transition, $N_h$ reflects the engine working state directly and $T_4$ guarantee the safety of the engine. Thus, it is reasonable to take the following objective function:

$$\min J = \omega_1 \int_{t_0}^{t_f} (N_h - N_{h_{\max}})^2 dt + \omega_2 \int_{t_0}^{t_f} (T_4 - T_{4_{\max}})^2 dt \tag{5.36}$$

where $x = (N_h, T_4)$ is set as the controlled variable and $u = (W_f, A_8)$ is set as the controlled variable; $t_o$ is the starting time of the acceleration and $t_f$ is the ending time. Then discrete the formula to get

$$\min J = \alpha_1 \left(1 - \frac{N_h(k)}{N_{h_{\max}}}\right)^2 \Delta t + \alpha_2 \left(1 - \frac{T_4(k)}{T_{4_{\max}}}\right)^2 \Delta t, k = 1, 2, \ldots n \tag{5.37}$$

where $\alpha_1, \alpha_2$ are the weight coefficients, respectively.

More specifically,

$$\begin{cases} \alpha_1 = \dfrac{(1-\frac{N_h(k)}{N_{h\max}})^2}{(1-\frac{N_h(k)}{N_{h\max}})^2+(1-\frac{T_4(k)}{T_{4\max}})^2} \\ \alpha_2 = \dfrac{(1-\frac{T_4(k)}{T_{4\max}})^2}{(1-\frac{N_h(k)}{N_{h\max}})^2+(1-\frac{T_4(k)}{T_{4\max}})^2} \end{cases} (k = 1, 2, ...n) \quad (5.38)$$

where: $\alpha_1+\alpha_2=1.\alpha_1 \geq 0.08, \alpha_2 \geq 0.08$; $\Delta t$ is step length and set it 0.02 s.

**Constraint Function**

The process of acceleration requires the physical limit of the shaft speed, the maximum temperature for turbine blades, the maximum limit of the burner (or combustor) pressure and the surge (or stall) limit of the compression system. Here we consider the following several limitations with the component-level model.

$$\begin{array}{ll} N_h \leq 12674 rad/\min, & N_l \leq 9907 rad/\min \\ T_4 \leq 1604k, & far \leq 0.022 \\ smc \geq 0.1, & smf \geq 0.1 \end{array}$$

Meanwhile physical constraints also exist as:

Flue flow limit: 0.3 kg/s $\leq W_f \leq$ 1.144 kg/s.

Nozzle area limit: 0.2642 m$^2$ $\leq A_8 \leq$ 0.2935 m$^2$.

The maximum rate of flue flow changing: $\Delta W_f \leq$ (0.6kg/s)/s.

The maximum rate of nozzle area changing: $\Delta A_8 \leq$ 0.2m$^2$/s.

### 5.3.3 Results and Analysis

Based on the existed engine model, the two optimization routines (ASM and SQP) in accordance with the objective function and the constraint function are developed. The MATLAB program implementation of the optimization is shown by different color as the following Fig. 5.12.

It is seen from Fig. 5.12 that SQP performs faster than ASM to get the expected engine working state from the labeled points. The specific time consumption is: SQP takes 2.46 s and ASM takes 2.54 s.

It is also seen that there is no significant difference between ASM and SQP for $A_8$. The $A_8$ optimization of SQP gradually decreases in the initial 0.5 s, and increases after the process. This is very close to the actual control plan; on the contrary, ASM remains constant. This is related to the optimization program realization. For $W_f$, the response from ASM is a bit slower than that from SQP and consumes less fuel

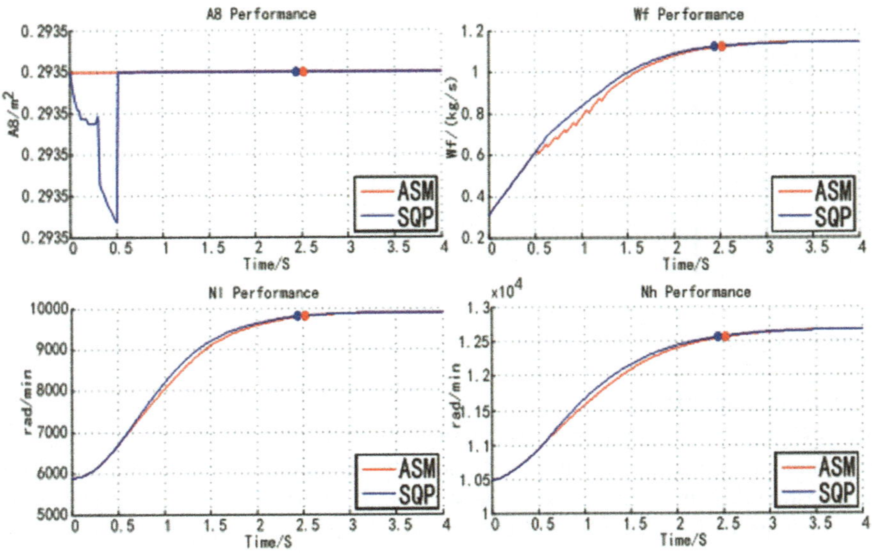


**Fig. 5.12** Simulation comparison of $A_8$ $W_f N_l$ and $N_h$

at the same time. From the comparison between $N_l$ and $N_h$, SQP responses faster than ASM, while tracking performance for both $N_l$ and $N_h$ is acceptable without steady-state error.

Figure 5.13 shows that optimization results for $T_4$, $smc$, $smf$, and $far$. It is seen that the response from ASM resembles slight oscillation between 0.5 s to 1.4 s. Turbine temperature from ASM is also smaller than that from SQP at the same time. Since $T_4$ has a strong correlation with the operational lifespan of aeroengine, it is expected that ASM may be able to provide life-extending control. Meanwhile, the surge margin of fan and compressor limit the engine operating normally hard, the values of surge margin get higher, the more reliability of engine operating. Likewise, it can be seen from the comparison between 0.5 and 1.4 s ASM maintains a relatively higher compressor (or fan) surge margin value than SQP. The same situation is also maintained at a lower $far$ which is more fuel-economic.

Finally, Fig. 5.14 shows the weighting values for $\alpha_1$, $\alpha_2$. It is obvious that strong oscillation exists for ASM while smooth transition for SQP.

### 5.3.4 *Conclusions for Comparative Study for Optimization-Based Designs*

A different perspective has been provided for optimization-based designs to optimize $W_f$ and $A_8$, with the weighted $N_h$ and $T_4$ as the objective function. Through comparison, some useful guidance can be extracted:

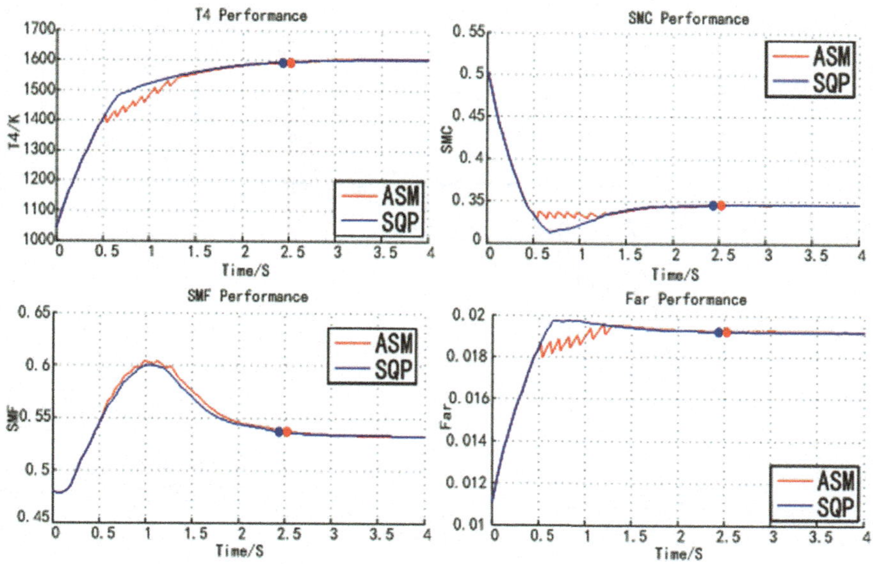


**Fig. 5.13** Comparison of $T_4smcsmf$ and $far$

(1) Fuel flow with oscillations as the optimization result of ASM, would require the fuel metering actuation being active, and thus reduces its operational lifespan.

(2) However, since the time of the acceleration is short and the transition process takes small proportion of the whole flight, it does not count much to evaluation index.

(3) Thus, it is reasonable to conclude (expect) that optimization-based control integration would not differentiate much among optimization routines that the designer utilizes.

## 5.4 Summary

Aeroengine transient control has been challenging due to its stringent requirements from both performance and safety. Many methodologies have been proposed such as conventional schedule-based methods, linear parameter varying, multi-objective optimization and evolutionary computations etc. These approaches have been well-established and led to a series of significant results. However, they are either not providing limit protection or requiring exhaustive computational resources, particularly when generating results into full flight envelope applications. Consequently, a compromise between limit protection and computational complexity is necessitated. This chapter considers a sequential quadratic programming-based method and active set method for full flight envelope investigations. The proposed methods

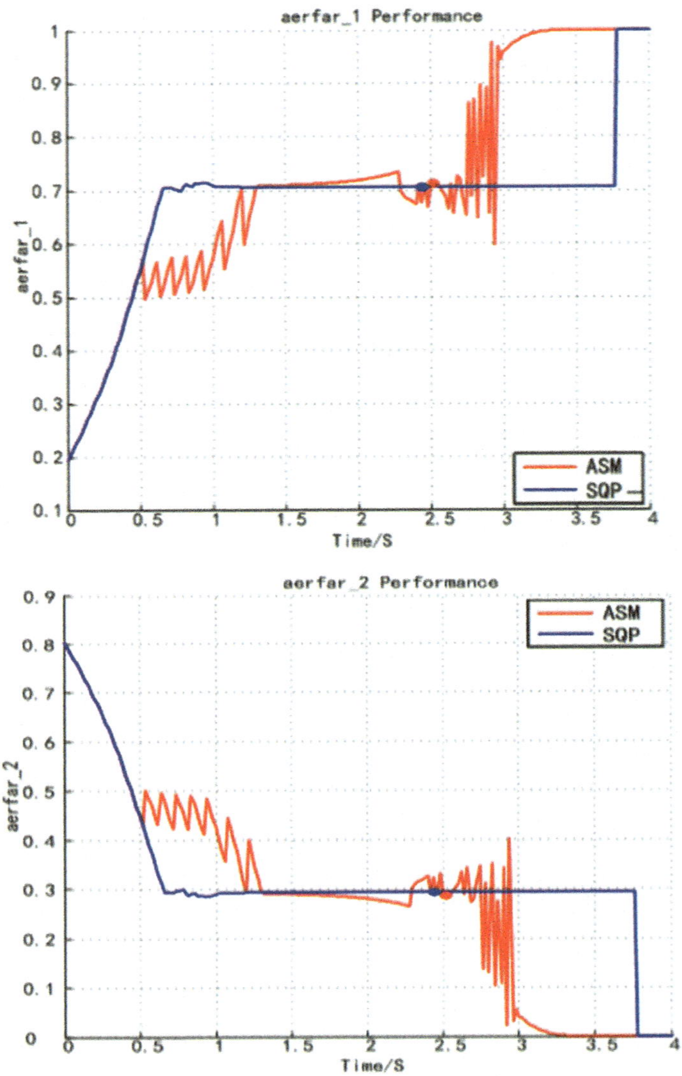


**Fig. 5.14** Comparison of $\alpha_1, \alpha_2$

can provide important design guidance and the corresponding claims are validated through detailed analysis and simulations.

# References


1. Jaw, L.C., and J.D. Mattingly. 2009. *Aircraft Engine Controls: Design, Systems Analysis, and Health Monitoring*. Reston, Virginia: AIAA Inc.
2. Lichtsinder, M., and Y. Levy. 2006. Jet engine model for control and real-time simulations. *Journal of Engineering for Gas Turbines and Power* 128: 745–753.
3. Bendtsen, J.D., J. Stoustrup, and Trangbak. K. 2003. Bumpless transfer between advanced controllers with application to power plant control. In *Proceedings of 42nd IEEE Conference on Decision and Control* 2003, pp. 2059–2064.
4. Zheng, K., T. Basar, and J. Bentsman. 2009. H∞ bumpless transfer under controller uncertainty. *IEEE Transactions on Automatic Control* 54 (7): 1718–1723.
5. Tumer, M., N. Aouf, D.G. Bates, I. Postlethwaite, and Boulet, B. 2002. A switching scheme for full-envelope control of a V/STOL aircraft using LQ bumpless transfer. In *Proceedings of 2002 IEEE International Conference on Control Applications*, pp. 120–125.
6. Rugh, W., and J. Shamma. 2000. Research on gain scheduling. *Automatica* 36: 1401–1425.
7. Leith, D., and W. Leithead. 2000. Survey of gain-scheduling analysis and design. *International Journal of Control* 73 (11): 1001–1025.
8. Apkarian, P., P. Gahinet, and G. Becker. 1995. Self-scheduled H∞ control of linear parameter-varying systems: A design example. *Automatica* 31: 1251–1261.
9. Wu, F., X. Yang, A. Packard, and G. Becker. 1996. Induced L2-norm control for LPV systems with bounded parameter variation rates. *International Journal of Robust and Nonlinear Control* 6: 983–998.
10. Stilwell, D., and W. Rugh. 2000. Stability preserving interpolation methods for the synthesis of gain scheduled controllers. *Automatica* 36: 665–671.
11. Chang, Y., and B. Rasmussen. 2008. Stable controller interpolation for LPV systems. In *Proceedings of ACC 2008*, pp. 3082–3087.
12. Balas, G. 2002. Linear parameter-varying control and its application to a turbofan engine. *International Journal of Robust and Nonlinear Control* 12 (9): 763–793.
13. Bruzelius, F. 2004. *Linear Parameter-varying Systems-an Approach to Gain Scheduling*. PhD Thesis, Chalmers University, Göteborg, Sweden.
14. Henrion, D., L. Reberga, J. Bernussou, and F. Vary. 2004. Linearization and identification of aircraft turbofan engine models. In *Proceedings of IFAC Symposium on Automatic Control in Aerospace*, St. Petersburg, Russia.
15. Reberga, L., D. Henrion, J. Bernussou, and F. Vary. 2005. LPV modeling of a turbofan engine. In *Proceedings of IFAC World Congress on Automatic Control in Aerospace*, Prague, Czech Republic.
16. Vary, F., and L. Reberga. 2005. Programming and computing tools for jet engine control design. In *Proceedings of IFAC World Congress on Automatic Control*, Prague, Czech Republic.
17. Gilbert, W., D. Henrion, J. Bernussou, and D. Boyer. 2010. Polynomial LPV synthesis applied to turbofan engines. *Control Engineering Practice* 18: 1077–1083.
18. Mayne, D.Q., J.B. Rawlings, C.V. Rao, and P.O.M. Scokaet. 2000. Constrained model predictive control: Stability and optimality. *Automatica* 36: 787–814.
19. Brunell, B.J., R.R. Bitmead, A.J. Connolly. 2002. Nonlinear model predictive control of an aircraft gas turbine engine. In *Proceedings of 41st IEEE Conference on Decision and Control*, Las Vegas, USA.
20. Fleming, P.J., and R.C. Purshouse. 2002. Evolutionary algorithms in control systems engineering: A survey. *Control Engineering Practice* 10 (11): 1223–1241.
21. Lyantsev, O.D., T.V. Breikin, G.G. Kulikov, and V.Y. Arkov. 2003. On-line performance optimization of aero engine control system. *Automatica* 39: 2115–2121.
22. Gill, P.E., M.W. Leonard, and L.O. Jay. 1998. An SQP method for the optimal control of large-scale dynamical systems. *Journal of Computational and Applied Mathematics* 120 (12): 197–213.

23. Maurer, H., and C. Buskens. 2000. SQP-methods for solving optimal control problems with control and state constraints: Adjoint variables, sensitivity analysis and real-time control. *Journal of Computational and Applied Mathematics* 12: 85–108.
24. Shi, Y., Q. Tu, Y. Cai, C. Qiu, and Z. Tan. 2013. Optimum design method of acceleration and deceleration control schedule for gas turbine engine. *Journal of Aerospace Power* 28 (11): 2567–2571.
25. P.M. Morse, and G.E. Kimball. 2012. *Methods of operations research.* Courier Corporation.
26. J.H. Horlock. 2013. *Advanced gas turbine cycles: A brief review of power generation thermodynamics.* Elsevier.
27. Støren, S., and T. Hertzberg. 1995. The sequential linear quadratic programming algorithm for solving dynamic optimization problems—A review. *Computers & Chemical Engineering* 19 (1): 495–500.
28. Krishnan, D., P. Lin, and A.M. Yip. 2007. A Primal-dual active-set method for non-negativity constrained total variation de-blurring problems. *IEEE Transactions on Image Processing* 16 (11): 2766–2777.
29. Zhao L, D. Fan, and W. Shan. 2008. A global optimization approach to aero-engine under transient conditions. *Journal of Aerospace Power* 04
30. Chipperfield, A.J., B. Bica, and P.J. Fleming. 2002. Fuzzy scheduling control of a gas turbine aero-engine: A multiobjective approach. *Industrial Electronics, IEEE Transactions on* 49 (3): 536–548.
31. Clark, D., M.R. Bache, and M.T. Whittaker. 2008. Shaped metal deposition of a nickel alloy for aero engine applications. *Journal of materials processing technology* 203 (1): 439–448.
32. Köller, U., R. Mönig, B. Küsters, et al. 1999. Development of advanced compressor airfoils for heavy-duty gas turbines: Part I—design and optimization. In *ASME 1999 International Gas Turbine and Aeroengine Congress and Exhibition. American Society of Mechanical Engineers, 1999: V001T03A021-V001T03A021.*
33. Zheng, H., H. Xu, and G. Zhu. 2008. Adaptive fuzzy control based on variable universe and its application to aero engine turbine power generator. *Control Theory & Applications* 2: 253–256.
34. Nettleton, D., and V. Torra. 2001. A comparison of active set method and genetic algorithm approaches for learning weighting vectors in some aggregation operators. *International journal of intelligent systems* 16 (9): 1069–1083.
35. Ceng, G., and D. Fan. 1999. Application of SQP to acceleration control of turbofan engine. *Journal of Aerospace Power* 03.
36. Xia, C., J. Wang, G. Shang, and M. Zhou. 2012. component-level modeling and analysis of aeroengine based on matlab/Simulink. *Aeroengine.*

# Chapter 6
# Conclusions and Further Developments

## 6.1 Conclusions

The concept of model-based design has been popularized over the past few years in Chinese aviation industry. Administrational and supervisional architectures have been designed to support the model-based design projects. Meanwhile the enabling technologies in smart sensors, high response actuators etc. have been developed into a readiness level up to 5–7, and it is expected that advanced controls including nonlinear control will be applied. In China, however, all of the civil and military engines still use conventional PI control methodology, and no essential research on advanced control is carried out. The development in control theory and control engineering has not been taken advantage of in the field of aeroengine controls. The results presented here, should shed some light on this important line of research.

Specifically, the research themes in the monograph provide a systematic investigation to nonlinear control of aeroengines. Gas turbine control has been challenging due to its both time-critical and safety–critical natures. For example, typical FADECs for implementing control algorithms are only 20 ms, which frustrates many advanced optimization-based control design approaches; meanwhile many limits exist such as pressures and temperatures in compressors, turbines and combustors etc., and this enforces stringent boundaries on control solutions, which can lead to catastrophic failures should these limits be violated. Consequently, conventional approaches to aeroengine controls are usually of trial-and-error nature, and this calls upon systematic and deep investigations into advanced control of gas turbine power systems in general.

This is attempted in this monograph with a project entitled "Model-based Nonlinear Control of Gas Turbine Power Systems (MBNC-GTPS)." This is the first project aiming to a systematic investigation on nonlinear control of aeroengines in China. The project has addressed a series of important problems including nonlinear real time modeling, model-based nonlinear control techniques, and integrated modeling and control over large flight envelope etc. As a consequence, a number of important results have been obtained:

 
J. Wang et al., *Model-based Nonlinear Control of Aeroengines*,
https://doi.org/10.1007/978-981-16-4453-5_6

(1) *Modeling of aeroengine systems*: both component-level and identification-based models have been extensively studied and compared; nonlinear real time modeling has been conducted including improved back-propagation neural networks, Hammerstein-Wiener representations, as well as nonlinear auto-regressive with exogenous input models.

(2) *Nonlinear set point control of aeroengines*: a generalized Gronwall-Bellman lemma approach has first been proposed for set point control of aeroengines, which can provide a tight bound on aeroengine transition speed over a specified area; a control Lyapunov function-based design methodology has then been introduced to provide an a priory stability guarantee; this has further been furnished with its application to detuning the conventional PID controls for guaranteed stability. Joint design with generalized Gronwall-Bellman lemma and Lyapunov function has been proposed to provide fast transition of aeroengine with stability. Then considering the fact that for two-spool engines, the low-pressure rotational speed is decoupled with high-pressure rotational speed, indicating that only one of the two variables can be controlled. Henceforth a restricted Lyapunov function approach has been proposed to tackle this situation. To further enhance the regulation time, a finite-time version of the restricted Lyapunov function method has also been proposed for significantly improve the set point control performance.

(3) *Nonlinear transient control of aeroengines*: different from the state-of-the-art approaches to model-based nonlinear transient control of aeroengines, a nonlinear generalized minimum variance-based aeroengine transient control has been proposed which allows the engine model to be practically any type, e.g. lookup tables, programming codes or any other non-analytical modules with channel delay. Essentially, a direct control of aeroengine component-level model has been allowed with a feature of easy design and detuning of controllers. Pushing this important line of investigation, this nonlinear generalized minimum variance control (NGMV) method has been developed a predictive counterpart, namely nonlinear predictive generalized minimum variance control (NPGMV). Then the integration of both to achieve robustness against disturbance has been introduced. This integration is novel in the literature since it adopts an anti-causal operator with an online tuning feature. Finally, to conquer the issue of sensor/actuator inertia, as well as addressing the fixed controller problem, a switching-based aeroengine transient control design method has also been presented. The last method, although rule-of-thumb, does provide a simple, economic, easy-to-detuning solution to improve the transient performance of aeroengines.

(4) *Optimization-based aeroengine nonlinear control integration*: the three basic functions of aeroengine controls consist of set point control, transient control and limit protection control. While set point control is relatively independent referring to aeroengine operating on steady state conditions. Transient control and limit protection control must be integrated due to the fact that it is during transition process that the engine approaches to its operational limits. To forge the integration of transient control and limit protection control, a

sequential quadratic programming-based method and active set method for full flight envelope control have been investigated. The proposed methods can provide important design guidance and the corresponding claims are validated through detailed analysis and simulations. However, it must be warned that all the optimization-based methodologies are inherently suboptimal due to discretization of performance index as well as utilization of linearized steady-state models during transition processes. Indeed, the engine dynamics varies significantly over the flight envelope and it is very difficult to obtain a mathematically analytical expression for the engine model. On the other hand, large dynamic uncertainty exists that prevents accurate models to be presented to control designers, in addition to the fast transient process with very limited computational power, e.g. there are only several seconds for engine to accelerate from idle to takeoff, while the corresponding control unit has a very slow computational speed, with a typical 15-20 ms sampling rate. Even for such limited computational resource, only 15–20% can be used for control and the remaining needs to be retained for monitoring engine key states to ensure reliability and safety. Thus, the optimization-based sub-optimal solutions achieve a balance between computational demanding and global optimality. In practice, however, such suboptimal solutions can indeed provide better acceleration properties than conventional schedule-based methods. As a consequence, it will be very useful for a methodology to provide design guidance. The proposed method does result in a series of results suggestive to practical implementations.

Consequently, a systematic design methodology has been developed which provides consistently enhanced performance over a large flight/operational envelope.

## 6.2 Further Developments

The monograph has covered from aeroengine modeling to nonlinear control. From system engineering perspective, next step of research will focus on the nonlinear control design methodology to be integrated with fault detection and the corresponding hardware-in-loop testing, rig test and even flight test, providing fuel economic demonstration. Meanwhile, with the new and fast development in model-based design and systems engineering, namely model-based system engineering (MBSE), a wide range of engineering industries begin to forge together to address new challenges. Towards this, several avenues can be suggested for further developments in the field of aeroengines.

### 6.2.1 From Modeling to Digital Twin

Modeling has played an important role practically during all the phases in aeroengine research and development process. It is also well aware that multiple types of models exist depending on different classification schemes. As list in Table 2.1, Chap. 2, in terms of model types, it can be classified as component level models and identification models; in terms of real time performance, there are real time models and non-real time models; in terms of model characteristics: linear models and nonlinear models; in terms of model property: dynamic models and steady state models etc. With research and development, new model types and methods will appear. These have been covered in Chap. 2 focusing on two basic modeling methodologies, namely first principle or physics-based modeling, and identification-based or data-based modeling.

The above models, both physics-based and identification-based models, are essentially "0-D models" in the sense that they are non-dimensional only considering the relationships among inputs, states, and output variables. Thus, to provide more details upon input-state-output evolutionary development over both space and time, 1-D, 2-D, or even 3-D models can be developed for aeroengine modeling. This forms one important line of further development in modeling. On the other hand, the simulation of aero-engine control system involves many disciplines due to its complex functions and architecture, which consists of mechanical, hydraulic, electrical, electronic system and so on. For each discipline, the modeling and simulation is usually dependent on different commercial software and tools, which makes the simulation, integration and verification of system-level model very difficult. To address this "multi-software, multi-platform" issue, a multi-domain co-simulation method is necessitated, e.g. based on Functional Mock-up Interface (FMI) standard to integrate models developed by different software or tools.

To give an example, Fig. 6.1 shows the structure diagram of a large bypass ratio turbofan engine digital control system. The digital electronic controller (EEC) is the central component of the digital control system, which mainly includes hardware circuit and software control logic. It receives the feedback signals from sensors and actuator, and outputs control command to ignition device, startup device and hydraulic machinery device after calculation. The fuel valve and exhaust valve of the engine are controlled by the hydraulic mechanical device to make sure that the engine can meet the requirements of the safety and performance of the aircraft. The typical control components of the system are selected to establish a multidisciplinary model, including rotor speed sensor, EEC conditioning circuit, fuel actuator, engine and EEC rising edge calculation module and control law model. To simulate the above components or functions:

(1) Simulink is chosen to build rotor speed sensor model, rising edge calculation module and control law model;
(2) Modelica is used to build rotor speed sensor conditioning circuit model;
(3) AMESim is utilized to build fuel actuator model;
(4) C language and Simulink to build engine mode; and finally

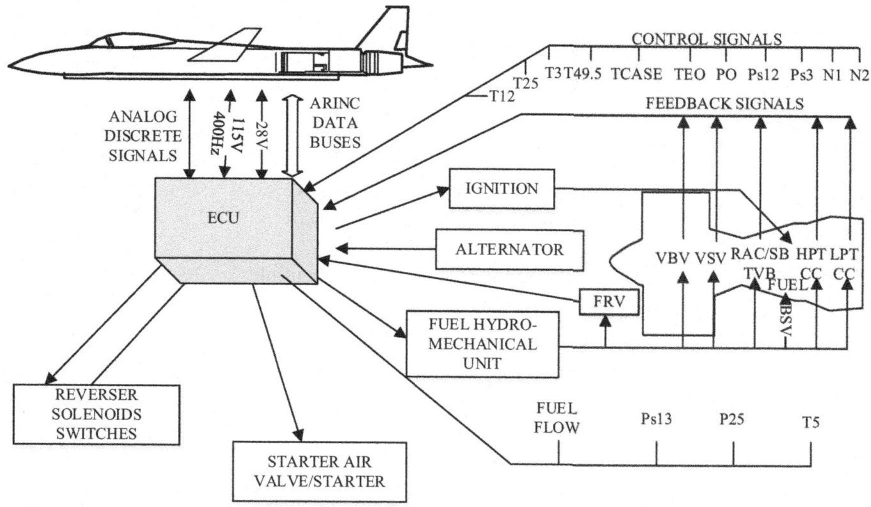


**Fig. 6.1** Control system diagram for a typical turbofan engine

(5) Integration of the above models and control law is completed based on FMI to realize co-simulation.

Examples of the above model types as modeled in each software environment are given in Fig. 6.2. The Integration of FMI-based co-simulation is shown in Fig. 6.3.

Finally, the simulation results are shown from Fig. 6.4 for full digital simulation. The start and stop stages are open-loop control, and the rest stages are closed-loop control. The actual speed is the feedback value of the engine model and the measured speed is the speed calculated from the rising edge of EEC. The relative error between them is shown in (b), with a maximum of no more than 0.4%. The source of the steady-state error is that there is an error of $\pm 1$ counting clock pulse when measuring the speed by the cycle measuring method. The relative error is $E = (1/m_0) \times 100\%$, and $m_0$ is the pulses number of internal clock in a period of the measured rotor speed. In simulation, the period of internal clock can be considered as simulation step size. If the simulation step size is set to $10^{-6}$ and the rotor speed is 1000 rpm, then the relative error is 0.1%. The simulation results comply with the theoretical analysis. The error of dynamic process is larger than steady state. This is because there will be a delay of pulse signal when measuring rotor speed by cycle measuring method, which will lead to larger error at the initial stage. (c) shows the effect of LVDT displacement closed-loop control. It can be seen from the figure that the actual displacement of the actuator can exactly track the demand displacement.

Co-simulation is thus another important methodology for multi-disciplinary model integration and cross-platform simulation, providing a baseline for high-fidelity control system design, simulation, integration, and testing. However, the co-simulation is still digital and does not solve the issue of "adaptive in real environment." That is, the model is not evolved with true aeroengine in operation. With

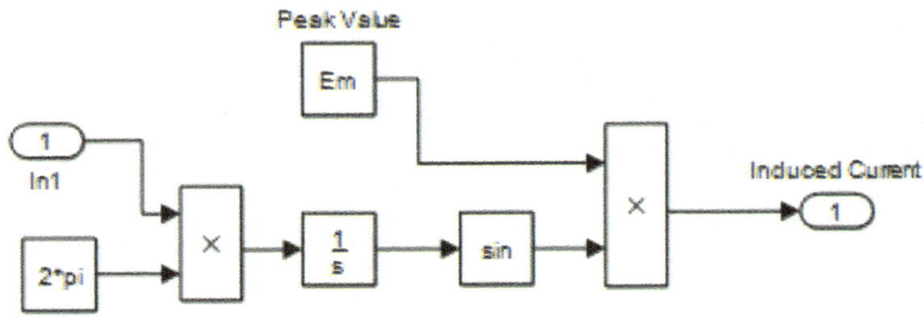


(a) Rotor speed sensor model in Simulink

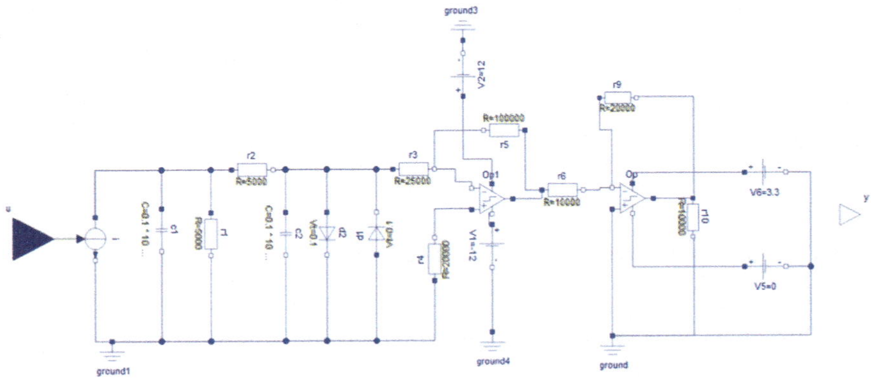


(b) Rotor speed sensor conditioning circuit model in Modelica

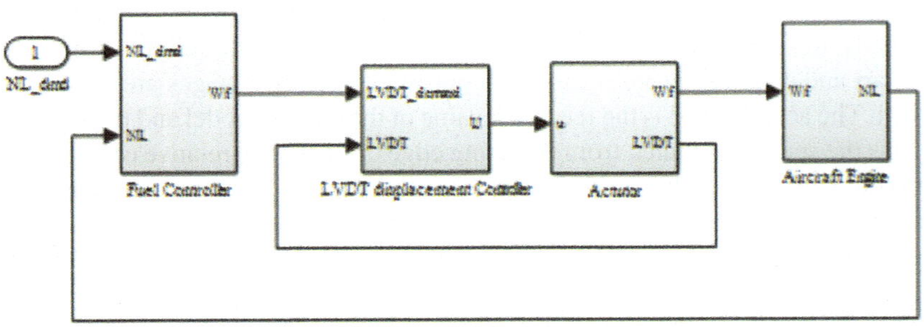


(c) Speed closed-loop control model in Simulink

**Fig. 6.2** Models in different software environment for each model

the development of "Internet of Things" or "Cyber-physical Systems," the next leap in modeling will be digital twin.

A digital twin is the generation or collection of digital data representing a physical object. It is essentially a software representation of a physical asset, system or process for optimization of business process. Henceforth digital twins are the outcome of continuous improvement in the creation of product design and engineering activities, from computer aided drafting/computer aided design (CAD) to model-based systems

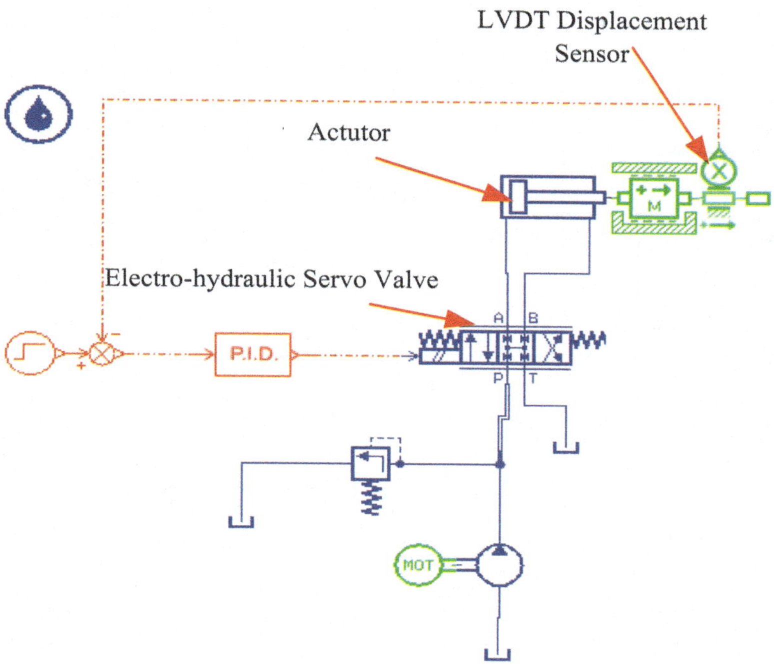


(d) Fuel metering model in AMESim

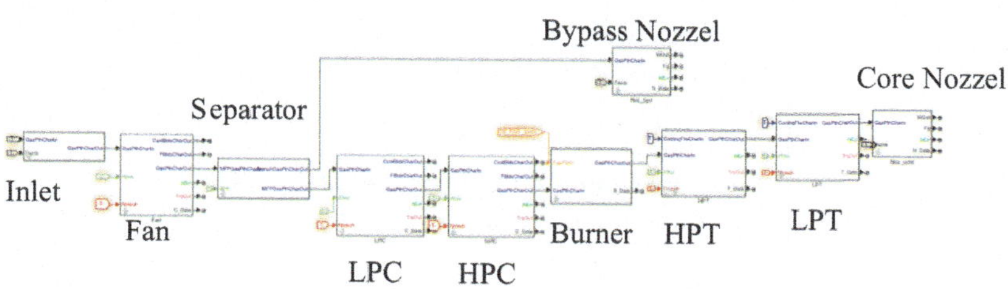


(e) Aerongine model using C/Simulink

**Fig. 6.2** (continued)

engineering (MBSE). Thus, modeling is increasing integrated with aeroengine health management to be covered in Sect. 6.2.3.

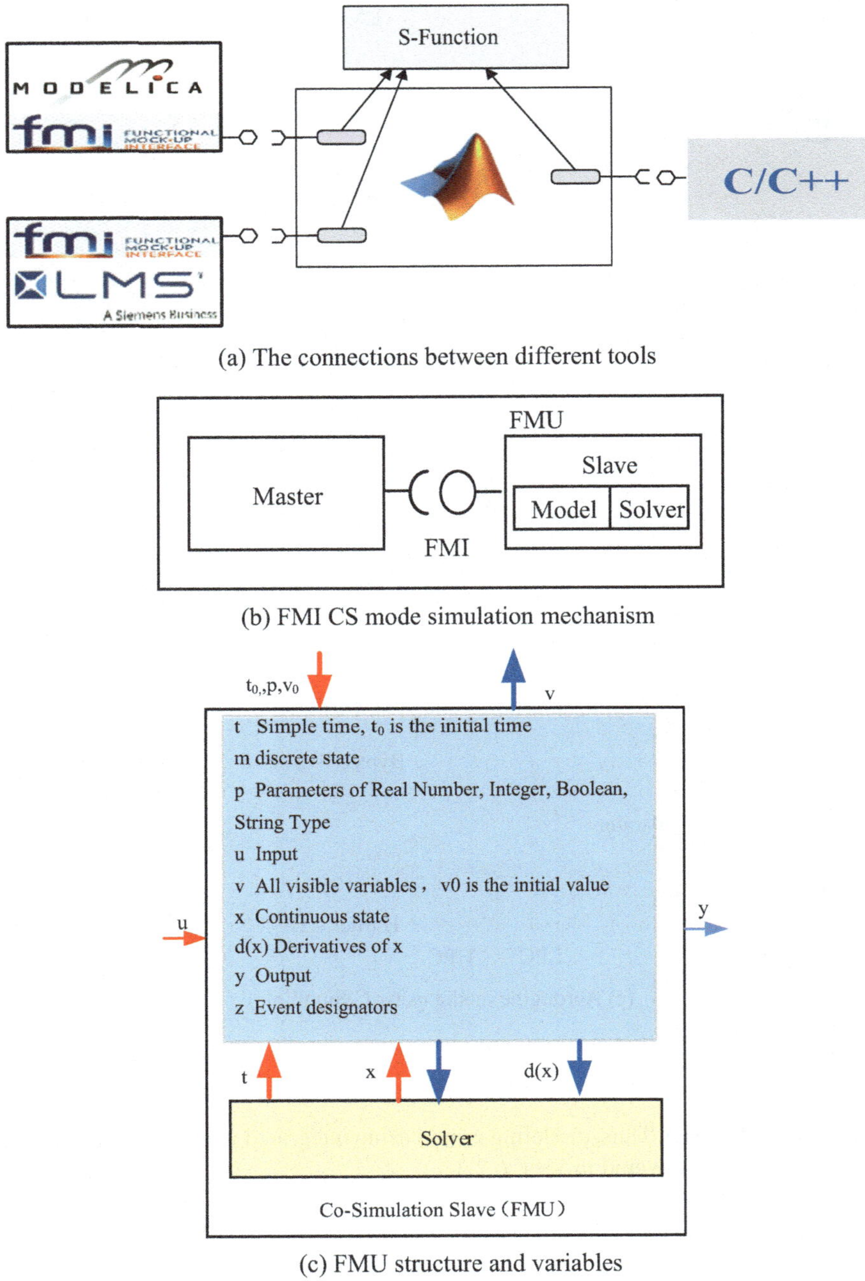


**Fig. 6.3** Integration of FMI-based co-simulation Framework

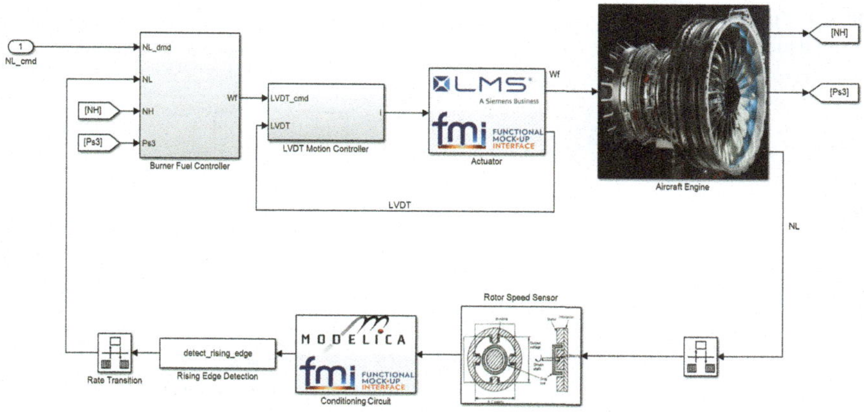


(d) Full-Digital co-simulation (Simulink Master)

**Fig. 6.3** (continued)

## 6.2.2 *From Optimal Control to Performance Limit Excavation*

Aeroengine controls have been divided functionally into three basic controls: set point control, transient control and limit protection. These have been discussed extensively in Chaps. 3 and 4. Optimization-based methods are brought into force for integrating transient control and limit protection are discussed in Chap. 5. These optimal control designs are powerful bringing new performance to aeroengine operation. Yet one of the new developments is to further excavate the aeroengine performance. This can be accomplished through increasing the performance of the consisting components, or by more advanced control design methodologies. This performance limit control has not been explicitly proposed in the literature but obviously is an important line of research, particularly considering the fact that the components performance has almost been pushed to the limits.

One of the potential strategies to improve the control performance is to take the frequency information into consideration. However, the current approaches are not targeting the performance defined over a frequency band. Even though the minimum variance type of square errors is defined at narrow band frequencies, simply synthesizing an optimal controller using optimal control methodology does not provide information on important problems such as the limit of performance at the desired frequency. In fact, the problem of determining limit of performance is an important issue that has unfortunately overlooked for many design approaches. For example, the famous mixed sensitivity method takes care of this issue by minimizing $H2/H\infty$ norm for both sensitivity $S$ and complementary sensitivity $T$. But the choice of weighting functions for both $S$ and $T$ is of rule-of-thumb nature, aiming to maintain the ideal shapes for the norms of sensitivity functions (e.g. $S$ small in low frequencies and $T$

**Fig. 6.4** Full digital co-simulation based on FMI

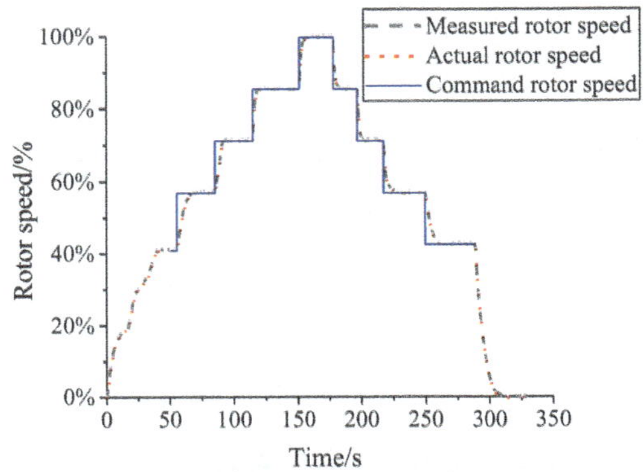


(a) Full-digital simulation of rotor speed control

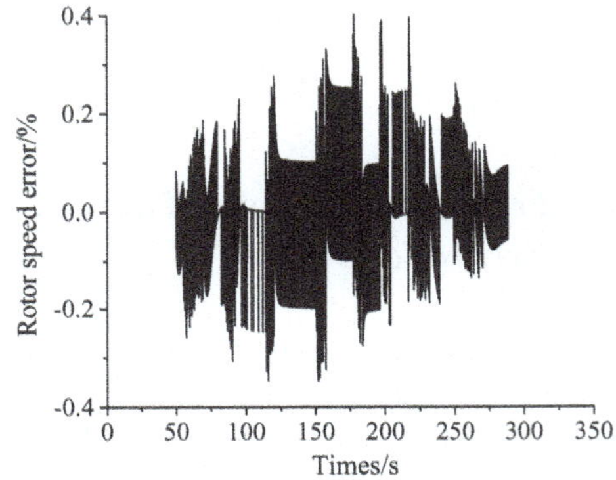


(b) Relative error between actual rotor speed and measured rotor speed

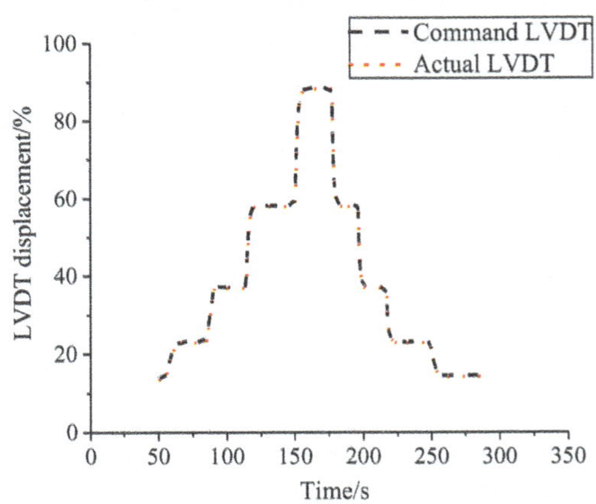


(c) Full-digital simulation LVDT displacement closed loop control small over high frequencies etc.). Important intuition on both solution existence and limit of performance at any discrete frequency $\omega_0$ or over a frequency band $[\omega_1, \omega_N]$ is thus lost. Indeed, it is claimed that the control design over a frequency band problem has been unsolved in the literature. Therefore, a methodology is necessitated that can possess the following features and capabilities *simultaneously*:

(1) Allow control design at any specific frequency or over any desired frequency band;
(2) Allow explicit determination of performance limitation for any frequency or over any desired frequency band;
(3) Allow performance improvement to current control.

In specific, feature (1) will provide a systematic sensitivity shaping method other than using weighting functions; feature (2) is even important, since prior to control design, a series of significant questions can be resolved, such as: (1) Does a controller exist that suppresses disturbance in both $S$ and $T$, and by how much? (2) Does a controller exist that will annihilate $S$ without enhancement of $T$? And similarly, (3) Does a controller exist that will annihilate $T$ without enhancement of S? Also (4) Can a controller be found that will suppress $S$ whenever $T$ is attenuated, and vise versa?

Feature (3) will be extremely important for the situation where the disturbance is known to be in a specific frequency band (but not necessarily being able to be modeled or even measurable). Then the proposed method can be utilized to further enhance the controller performance over the desired frequency band, while accepting performance deterioration outside that frequency band (due to Bode's integral relationships or waterbed effect). In fact, no control design methodology is capable of exploiting performance limitation over a specific and explicit frequency band, but it can be demonstrated that the following proposal can always improve the current control design, be it $H2/H\infty$, PID, or adaptive control etc. over an explicitly defined frequency band $[\omega_1, \omega_N]$ . In this sense, the proposed method is claimed to be UNIVERSAL to performance improvement at any discrete frequency or over any desired frequency band.

This method is coined as "Geometric Design" which nicely possesses all the features and capabilities. The concept is illustrated in Fig. 6.5, with capabilities demonstrated in Fig. 6.6 and Fig. 6.7. The methodology is also proposed by the authors of the monograph and the details can be referred in the literature easily. It is believed that this is an important attempt to aeroengine performance limit excavation, while expecting more design methods emerging to handle such important control design issues. After all, the objective of control design is to excavate the potential of systems by taking every piece of information into consideration.

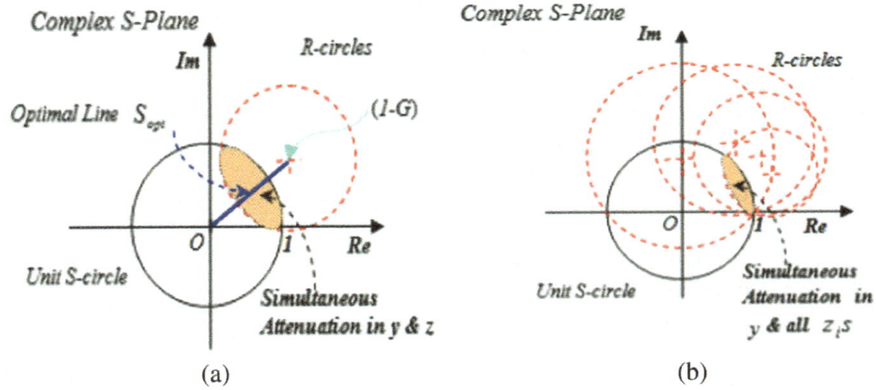


**Fig. 6.5** Geometry of $S$ and $R$ revealing fundamental performance limitation at a discrete frequency

### 6.2.3 *From Fault-Tolerance Control to Predictive Maintenance*

With rapid development of information technology, engineering systems have become more and more large-scale, inter-connected, and increasing-intelligent. However, in the field of maintenance and support of aviation equipments, the after-maintenance of traditional equipment after failure and the planned-maintenance according to the predetermined period have serious shortcomings in the ability to respond to rapidly changing new situations. These conventional maintenance methods not only consume resources and remain inefficient, but also have high costs. Comprehensive data from the US military shows that in the life cycle cost of weapons and equipment, the use and support costs account for 72% of the total cost. In the aerospace field, statistics conducted by the United States during the development of the X-34 and X-37 launch vehicles (Reusable Launch Vehicles, RLV) show that in order to ensure the success of the mission, the preventive measures carried out on the space shuttle during each mission period repair work costs 4 million U.S. dollars and requires about 200 maintenance personnel. Traditional maintenance methods have made the aerospace, defense and other industrial sectors suffer huge financial, material and manpower expenditures. Therefore, the defense acquisition reforms carried out by the United States since the 1990s and subsequent related large-scale project research have listed "affordability" as one of the pivotal considerations.

Different from the above two maintenance strategies, condition-based maintenance (CBM) is a strategy based on the actual state and development trend of the equipment. The basic idea is based on correct judgment and reliable prediction of the current and future state of the equipment. Condition based maintenance has significant advantages such as small scale, high efficiency, good affordability, and avoidance of major catastrophic accidents, so it is very suitable for maintenance support of large and complex systems. The description of the current state of the equipment, as well as the prediction of the state and failure of the next period of time, are the basis for

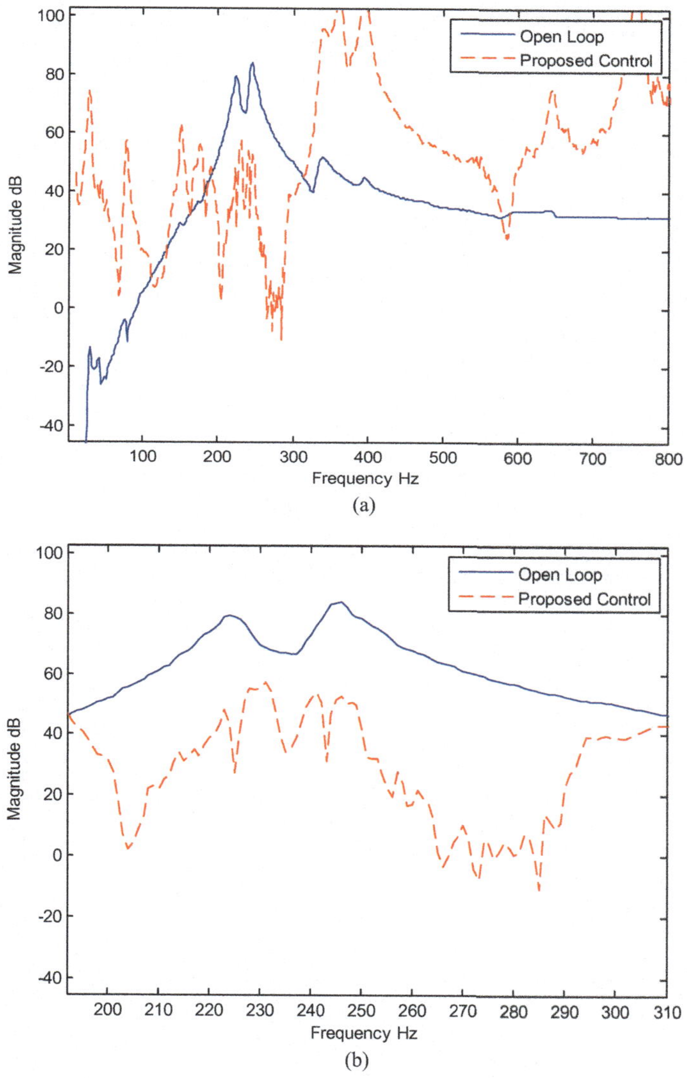


**Fig. 6.6** In-frequency-band performance limit excavation. **a** Proposed controller performance. **b** magnified view over frequency band [180, 310] Hz. Vibration suppression is achieved over this frequency band covering both the first and the second bending modes

realizing condition-based maintenance. Based on this demand, the concept of prognostic health management was first proposed in the aviation industry. In prognostic health management, prediction is the enabling technology for its implementation, referring to the use of sensors to obtain data information of the system, the help of various intelligent models and algorithms to evaluate the health of the system itself, the prediction before system failures, and thus providing a series of maintenance

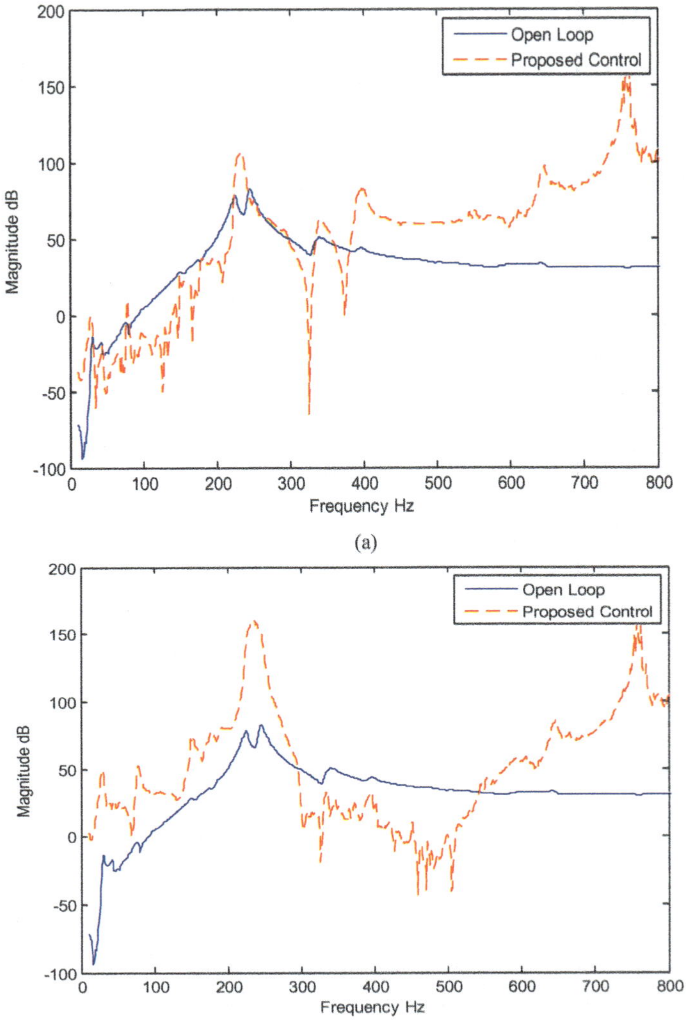


**Fig. 6.7** Out-frequency-band performance limit excavation—performance control performance improvement over. **a** Low Frequency. **b** High Frequency

support recommendations based on the available information resources. Figure 6.8 for a illustration.

Today the majority of aeroengine trending and prognostic functions are performed on the ground, where the in-flight data and pilot reports are transmitted from the onboard system—*real-time in-flight prognostics are still in development*. But efforts have been extensively made to integrate the data information and take full advantage of data mining for decisive maintenance support (Fig. 6.9).

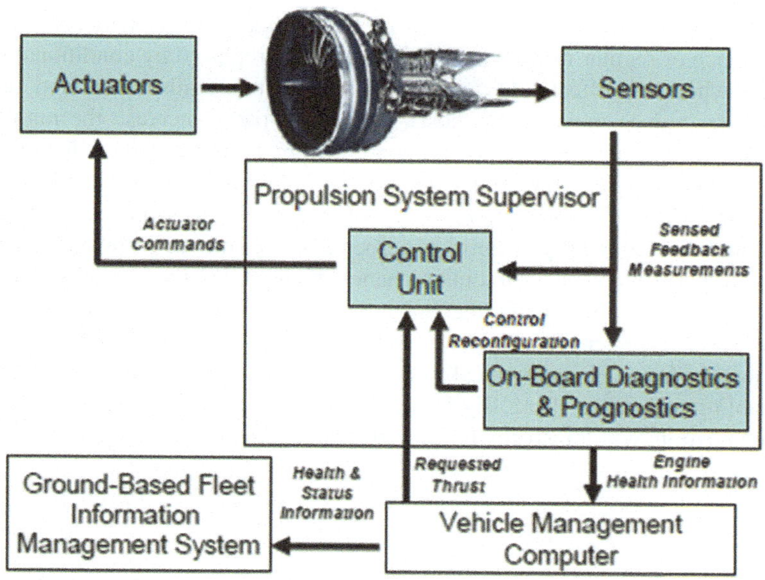


**Fig. 6.8** An illustration of aeroengine prognostic health management

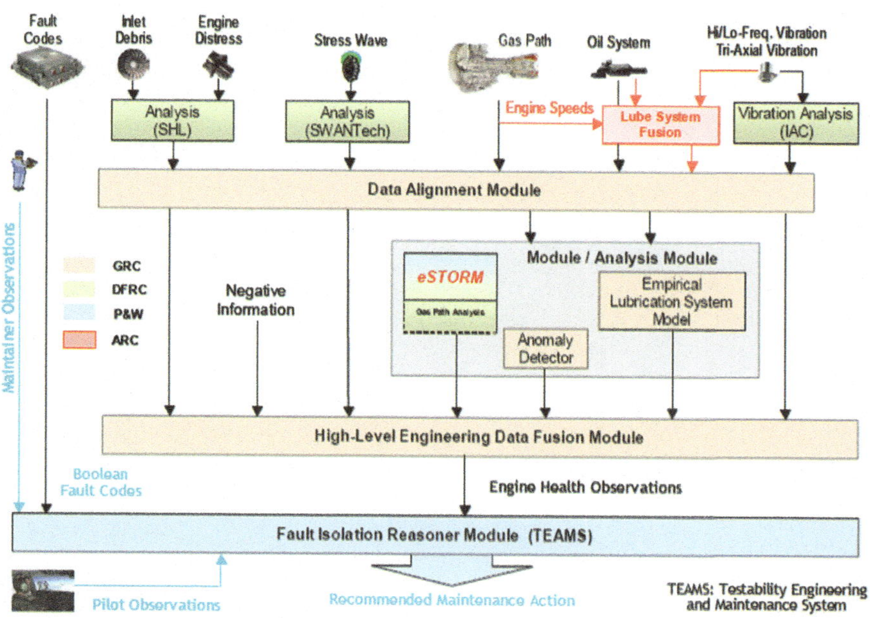


**Fig. 6.9** Data fusion for maintenance management

Thus, it is seen clearly that aviation prognostic technology enables after-event maintenance or regular maintenance strategies to be replaced by conditional maintenance, which can effectively improve aviation system maintenance and support capabilities, reduce maintenance costs, save spare parts, increase the number of dispatches, and achieve autonomous support, thereby fundamentally reducing maintenance support costs. Aeroengine monitoring and diagnosis system is increasingly integrated with control systems, providing better performance, safety, operability, and maintainability. Control of aeroengines has moved from fault tolerance control, to performance-oriented maintenance, finally targeting predictive management. This surely necessitates the integration of digital twin, advanced control, and health management for total asset and process optimization to deliver business performance. It can be optimistically foreseen that the developments together with the enabling technology from Big Data, Artificial Intelligence, and even Quantum Communication, will transform the century-old aviation industry to a new horizon.